AF430908

A Tale of
Two Twins

The Langevin Experiment of a Traveler to a Star

A Tale of
Two Twins

The Langevin Experiment of a Traveler to a Star

Lucien Gilles Benguigui

Technion – Israel Institute of Technology, Israel

World Scientific

NEW JERSEY · LONDON · SINGAPORE · BEIJING · SHANGHAI · HONG KONG · TAIPEI · CHENNAI · TOKYO

Published by

World Scientific Publishing Co. Pte. Ltd.
5 Toh Tuck Link, Singapore 596224
USA office: 27 Warren Street, Suite 401-402, Hackensack, NJ 07601
UK office: 57 Shelton Street, Covent Garden, London WC2H 9HE

Library of Congress Control Number: 2020946408

British Library Cataloguing-in-Publication Data
A catalogue record for this book is available from the British Library.

A TALE OF TWO TWINS
The Langevin Experiment of a Traveler to a Star

ISBN 978-981-121-909-2 (hardcover)
ISBN 978-981-121-910-8 (ebook for institutions)
ISBN 978-981-121-911-5 (ebook for individuals)

For any available supplementary material, please visit
https://www.worldscientific.com/worldscibooks/10.1142/11791#t=suppl

Typeset by Stallion Press
Email: enquiries@stallionpress.com

Preface

Is it really possible that after more than 100 years, something new may be published about this unbelievable result of the Theory of Relativity of a traveler who, after reaching a star, comes back younger than people on earth?

Today, it is almost impossible to publish an article on this subject in a respectable physics journal: the subject is exhausted. About ten years ago, I began to be interested in the subject and discovered two facts. The first is that this story (implying innumerable publications) did not receive a complete survey. The story is well known by physicists and people interested in Relativity, but I did not find an article recalling all of what happened from 1911 when the French physicist Paul Langevin published his article. I confess that, due to the huge expanse of the literature, I might have missed such an article, but the probability is small. The second fact is the almost infinite number of publications presenting a solution, explaining how the Theory of Relativity reaches this result for the younger traveler.

I wrote an article to fill the absence of a survey and of a standard solution accepted by all. Of course, I did not try a physics journal but I published in arXiv. However, the impact was very small and few authors mentioned my article.

After a time of frustration, I decided to write a book and looked for a publisher. I want to thank *World Scientific Publishing* for accepting my manuscript. For a long time, the subject was at the center of discussions and polemics, and I hope that my book will not awake new antagonisms. There is a little risk since in some parts of the book, I do not agree with well accepted opinions. I ask the reader to be careful in his reading before raising objections.

Finally, I want to acknowledge several important discussions with my colleague Ori Amos. All my family (my wife Suzanne, my children and my grandchildren) had been supportive and encouraged me to go on when I was hesitating. I express my thanks for all and in particular my granddaughter Hadar for her help in using Word.

Haifa, May 2020   Lucien Gilles Benguigui

Contents

Part III: Discussions   117

Chapter 1

Introduction

The thought experiment of Langevin is very well known and may be described in a few words. For a long time, this experiment had famous names such as the Twin paradox and the Clock paradox. A traveler travels to a star located at a distance of some light years from the earth with a velocity close to the light velocity. When he comes back to earth, his clock indicates that he traveled two years, but on earth, the clocks indicate that 200 years have passed.

From our daily concept of time, the story seems incredible, but this story is the result of the Theory of Relativity. In the huge literature on this subject, one finds different formulations of the paradox, but one does not need them, because the story by itself is very simple. The basic question of the paradox is: how is it possible that the traveler's clock and the earth's clocks do not indicate the same time? One may call this situation a "paradox", since in the Oxford Dictionary it is defined as

> A paradox is a seemingly absurd or self-contradictory statement in logic that, superficially, cannot be true but also cannot be false.

The first time that I heard about the Langevin experiment, my first reaction was that it is not possible since the two observers are in a reciprocal situation. It seems that Special Relativity and reciprocity are always associated in the minds of those who study it.

Today I do not teach but I thought to study some aspects of the history of physics. I began with the entropy and wrote an article. It was a pleasant exercise and I decided to continue to read more.

The history of Relativity attracted my attention and I remembered the Langevin experiment. I decided to look for an explanation.

When I began to think about the problem, I realized that in what I had learned as a student and in what I had taught to my students, there was no tool to explain this paradox.

The first thing to do is to understand well the paradox. Why do we expect that all the clocks will have the same indication of time at the end of the travel? The first and most evident reason is that we consider time to be a universal concept independent of the events in the world. In the present case, there is a perfect symmetry between the two observers such that the times cannot be different.

We shall begin with a detailed analysis of the travel from the point of view of the two observers. An engine first moves the traveler: a force is applied and it results in an accelerated motion. After attaining some velocity, the traveler cuts off the engine and moves at a constant velocity. Approaching the star, he activates his engine and a force is applied in the reverse direction. His motion is deaccelerated until he reaches the star with zero velocity. The return travel is the same but in the reverse direction. This is what an observer on earth sees, and we shall call him the stay at home.

Now we consider the traveler and ask: What does he see during the travel? From the point of view of the traveler, he is considered to be at rest. Whereas for the stay at home, the traveler begins his motion toward the star in the direction $x > 0$, the traveler sees the stay at home moving in the direction of negative x. First, he sees an accelerated motion of the stay at home, followed by a period of constant velocity and finally a deaccelerated motion until the stay at home becomes motionless. The two observers see the same motion, relatively one to other. Apparently, there is a perfect symmetry and consequently one expects equality of the clocks' indications. Langevin used an imagined language preferring to speak of the ages of the observers instead of their clocks.

However, the Theory of Relativity states that the clocks (or the ages) of the observers are not equal, and the traveler is younger. The problem is to find the mechanism of the symmetry breaking.

One finds very frequently that the difference between the two observers lies in the fact that the traveler is accelerated by his engine.

It is true. However, this argument is weakened because one supposes that the time when the traveler engine is at work is very short. In other words, the acceleration is present for so short a time that, apparently, one can neglect it. But a question cannot be avoided: Is it right to say that a very short acceleration time does not influence the result of the journey? For many people, the answer is yes, but they say so without checking if it is right.

However, some scientists insist in saying that the traveler feels the acceleration but not the stay at home. This is also true, but the theory does not take into account the feeling of the observers. One remains with the symmetry.

The question, for what reason are the times not equal, has received an incredible number of answers, with a large number of solutions. The subject is very exciting and one finds articles in not only physics journals but also journals of history of science, journals of philosophy of science, popular scientific journals, and even in literature. R.B. Heinlein (1956) wrote a science fiction novel (Time for the Stars) where identical twins are important personages in the plot. Philosophers have felt that the new "time", which should be equal in all circumstances, requires new thinking.

When I began to read several textbooks, I was awestruck by the variety of solutions and it seemed to me that there are several errors. I felt that I am in a jungle. A jungle is a dense forest in which you have to find your way by yourself. It was exactly my impression. And when I continued my research in journals and reviews, this impression became stronger. I found long discussions, for example, in the book of Arzeliès (1966) and also other authors. There were (and I suppose still are) many repetitions of the same ideas in different journals. In short, I did not know what to do until I found one solution that gives in a very simple way the explanation for the age difference. Curiously, this solution was not frequently quoted in the textbooks, although it was published in 1952.

It appears to me that the entire story is a very peculiar phenomenon. I do not know of another problem in physics that has been the subject of publications and discussions from the beginning of the 20th century until today. The large majority of publications aimed at solving a problem for which the solution is already known!

The exact number of publications in the world is not possible to determine, because the Langevin experiment is included in the textbooks and journals in a large number of languages. In English, one can evaluate the number as several thousands. I thought that it is worthwhile to present the phenomenon and to try to find an interpretation for this special situation. The fact that solutions and discussions have appeared very regularly during the second part of the preceding century (some of which aimed at showing that the observers have the same age) is very intriguing and may be an indication of a special attraction of this problem.

The book is divided into three parts. It is evident that the reader must have direct access to the article of Langevin. Generally, one wants to remember only the story of the two observers, one staying on earth and one traveling to a star. However, the article is richer in detail than this little story, and in particular, it gives a clue for understanding the paradox. This is why I began with the translation of the Langevin article: it is now possible to access the English translation directly online.

In the second part, I start by presenting a few solutions, which I chose because they were the first solutions proposed by Lorentz, Langevin, Einstein, and Born. An essential solution of this part is the solution of Møller (1952), which is a simple solution using the Special Theory of Relativity.

I tried to also present qualitative explanations. By qualitative explanation, I mean explanations that do not need equations in order to be understood. In general, equations play the role of explanations but I think that a qualitative interpretation may be more convincing than equations. The remarkable situation of the Clock paradox is that the equations necessary to solve the problem may be obtained through the Special Theory of Relativity but when one wants to obtain qualitative explanations one needs the General Theory of Relativity. In fact, this is a means to close the debate on which theory is best to solve the problem.

The bibliography of the book is not very large. Because of the vast number of publications, I had to make a choice. Therefore, I chose the works that are necessary to follow the developments of this book. It is very likely that I missed several interesting articles.

In the third part, some famous debates and discussions are recalled. A special chapter is devoted to some tentative interpretations. There are so many questions that one may ask and surprisingly I did not find answers. The Clock paradox is unique in the history of physics and it is time to begin a detailed analysis. I think that one could discover many new things about physics and psychology.

I hope that this book will be of interest to two kinds of readers: first the physicists for whom I present a detailed analysis of a simple and realistic solution and also a qualitative interpretation of the age difference. Furthermore, historians of science and philosophers of science will find matter for reflection about a not so well-known phenomenon.

PART I

LANGEVIN

Chapter 2

The Story in Short

The story began with a lecture and an article by the French physicist
Paul Langevin in 1911. A lecture was given by Langevin at *Le congrés
interntionale de philosophie* at Bologna in April 1911, followed by the
publication of an article by him in *Scientia*, a journal that was edited
in Bologna from 1910 to 1988 under this name. *Scientia* published
articles in several languages, including French. Today, the text of the
article is available in *Wikisource* in French and also in English. One
can also access the original article online.

Paul Langevin was a remarkable French physicist of the 20th
century (1878–1946). He made numerous contributions to several
domains of physics. His works concern fundamental physics of
magnetism and the Brownian phenomenon. One important equation
concerning fluctuations bears his name. During World War I, he
developed devices for the detection of submarines, first with a
condenser and then with piezoelectric quartz, following the advice
of Pierre Curie. He followed closely the developments of the Theory
of Relativity and was one of the first physicists who contributed to
the diffusion of the theory in France.

2.1. The Langevin Article

In the article of Langevin, all that remains in all our memories is
the end. As is well known and as already mentioned, he imagined a
traveler leaving the earth toward a star with a velocity a little smaller
than the light velocity and coming back much younger than people
on the earth. Even for people having some knowledge of the Special

Theory of Relativity, there is something strange. As we saw above, apparently the two observers are at symmetrical positions. How is it possible that at the final encounter they do not have the same age?

It is clear that Langevin wanted to make a strong impression on philosophers and on people interested in the Theory of Relativity. He took a kind of extreme situation in which the concept of time is completely renewed and is very different from its habitual meaning. Since one of the novelties of the Theory of Relativity is that there is no universal time, the example of a traveler coming back younger than the people on the earth was very well chosen.

The Langevin article made a real impact on the interpretation of the Theory of Relativity. We shall see that in the Langevin article there is no paradox and no twins. Following During (2014), the word "paradox" appeared (maybe for the first time) in Max von Laue in 1923. The word "twins" appeared in the book of Weyl in 1922 may be for the first time. In order to give more strength to the problem, Weyl imagined that the two observers are twins, i.e., two men (or two women or a man and a woman as in some articles) with exactly the same age. Hence, the Langevin experiment received the name "Clock paradox" or "Twin paradox". Although there is no paradox (it is the conclusion of the theory), I shall call this thought experiment a "paradox" for the sake of simplicity.

As mentioned before, Langevin did not use the word "paradox" since for him it is only a consequence of the theory and as said by Einstein himself, it is a "curious" phenomenon. One has to note that in his article, Langevin does not explain how the age difference may be deduced from the Theory of Relativity. He indicates only that the source of the age difference is the acceleration when the traveler changes its direction and comes back to earth. This has important consequences; since the publication of the article, there has been a perpetual search for an explication.

But at the same time, this incredible result from the Theory of Relativity also became an argument against the theory (by the way, today there are always skeptics of the Theory of Relativity and the story of the traveler is one of their "strong" arguments). This can be seen from the article published in 1918 by Einstein in which there is a dialogue (which was an invention of Einstein) between a critic of the Theory of Relativity and a relativist. The first question presented by the critic is the problem of the paradox. The relativist (Einstein) gives his solution. This shows that very early, the Clock paradox was difficult to accept.

2.2. Einstein in Paris: The Meeting with Bergson

After World War I, Langevin worked for the reconciliation of French and German physicists. In 1922, he invited Einstein to Paris for lectures at the *College de France* and also for a meeting of Einstein with scientists and philosophers (Nordman, 2011).

The role of Langevin in organizing the visit of Einstein to Paris can be easily understood if we recall that he was well acquainted with Einstein and also had an interest in philosophy and social sciences (Bensaude, 1988). During his life, Langevin published (essentially in French) articles on various non-scientific subjects such as *L'csprit dc l'enseignement scientifique* (The spirit of the teaching of science), *Science et Laïcite* (Science and secularism), *Fascisme et civilisation*, and *Culture et humanités*. This was in the French tradition of associating science and philosophical reflection as Poincaré did and thus it is no surprise that the last meeting of Einstein in Paris was a meeting of scientists and philosophers.

This meeting took place at *La société française de philosophie* on 6 April 1922. There were present at this meeting philosophers (Brunschvicg, Bergson, Le Roy, and Meyerson), mathematicians (Hadamar, Cartan, Painlevé, and Levy), and physicists (Becquerel, Langevin, and Perrin). If it is memorable, this is because of the intervention of Bergson. When other participants made general remarks about some points of the theory, Bergson chose to present a philosophical analysis of the concept of time. He questioned the meaning of different times that appear in the Special Theory of Relativity and tried to show that the concept of time of physicists is not exactly the concept of time generally accepted (we shall come back to this problem later). The answer of Einstein cannot be called by any other word than "rude": he answered that he knows of only the time of physicists and the psychological time. In brief, for him, what philosophers can say about time was without any interest.

This rebuttal of Einstein did not prevent Bergson from publishing a book on the Theory of the Relativity shortly after his meeting with Einstein: "*Durée et Silmutanéité*" (*Duration and Simultaneity*). This book was vividly criticized by physicists and a controversy began with some French scientists. A very well-known and esteemed philosopher who writes on a physics theory is something exceptional. Until today, it has been difficult to understand what were the exact intentions of Bergson. A recent re-edition of *Durée et Silmultanéité* was published recently with discussions and analysis by philosophers.

The critic was very easy to handle for physicists until Bergson wrote that he did not agree with the accepted solution to the age difference in the Langevin experiment. An interesting point is that physicists and Bergson stood on their positions and neither was convinced by the arguments of the other. We shall analyze this controversy in a following chapter.

2.3. The First Reactions

As mentioned above, the Langevin experiment was largely commented on by physicists and, very early, several explanations were given (Lorentz 1914, Einstein 1918, Born 1922). From solutions of the problem or explanations, one can understand how the application of the Theory of Relativity yields the result given by Langevin: the stay on earth is older that the traveler. These solutions are very important, because all of them imply the role of acceleration.

However, in a very large number of articles, it was supposed that the times of accelerated motions of the traveler are so short that there is no necessity to take them into account. In other words, there is no need to introduce these accelerated motions into the analysis of the problem. Although Einstein in his 1918 paper indicates clearly that there is no necessity to suppose short times for the acceleration, this elimination of the acceleration became an integral part of the solutions. Only a few papers and a few textbooks suppose an intrinsic time for the accelerated portions of the travel.

2.4. The Deluge of Solutions After 1950

Until 1950, relatively few articles appeared on the subject, some of them denying the age difference, and even in these cases it does seem that this induces polemical discussions (Von Laue, 1923). For unclear reasons, after the fifties of the preceding century, the story took a completely different aspect. Instead of being a relatively calm subject, it became very "hot". There were three novelties: first, a complete solution was published by Møller in 1952, secondly, vigorous controversies took place with the physicists Dingle and Sachs, with the publication of several dozens of articles, and finally, there was a real deluge of solutions. From 1950 until today, a large number of articles have been published. This number is not very clearly known; some researchers estimate it at 25,000! But several thousands seem more reasonable.

The word "deluge" is justified when one sees the number of publications. We tried to count articles in some important physics journals (*Physical Review, Physical Review Letters, American Journal of Physics, Foundation of Physics, European Journal of Physics, Nature,* and *International Journal of Theoretical Physics*) and got the results shown in Fig. 2.1. We do not claim to have captured all

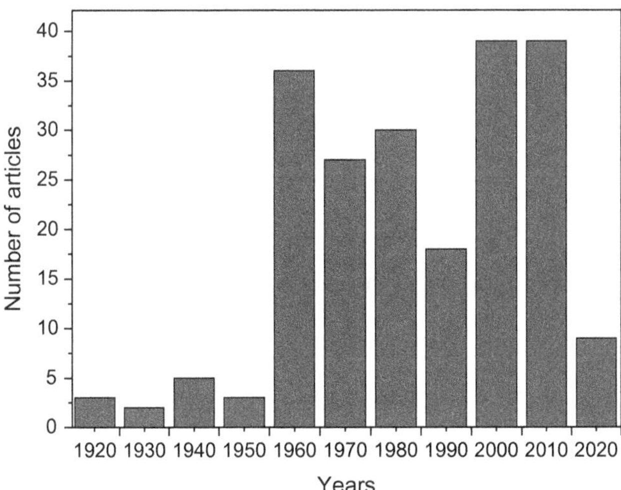

Fig. 2.1. The number of articles on the Clock paradox in physics journals. There have been 211 articles from the beginning. Note the change after 1950.

the articles, but the trend that emerges is clear. However, the reason for this avalanche of publications is unclear and we shall discuss some plausible explanations later.

The consequence of this abundance of articles is that today, it is almost impossible to publish an article on this subject in a good journal of physics. I suppose that for the editors, the subject has been exhausted and in fact, the question is: What may be said more on the Clock paradox? But the Internet is always there and there is no problem in publishing through this medium. This is the reason why one can find, from time to time, articles in various Internet sites. In particular, the site arXiv is very popular.

2.5. The Solution of Møller (1952)

Møller was among the few who dealt with the Clock paradox and abandoned the assumption of very short acceleration times. In fact, he follows closely the scheme proposed by Einstein in his 1918 article. He supposes that a force F is applied for some time until the velocity of the traveler is large enough, then the force is made null (and the velocity is constant), and finally the same force is applied but in a reverse direction until the velocity is zero. We shall detail the solution later, but one can already say that the point of view of the observer on the earth is deduced from the Special Relativity, but for the traveler, the General Relativity is applied.

The surprising thing is that this work of Møller was not very well known. His book is frequently quoted in the literature, but his specific solution is ignored. As we shall show below, the majority of articles on this subject continue to adopt the point of view of very short acceleration times and this is one of the reasons for the controversies about the Clock paradox. This neglect (whether deliberate or not) is puzzling.

2.6. The Dingle and Sachs Controversies

Herbert Dingle was very well known because of his opinions about Relativity. One can say that he was a reputable physicist with good esteem. He wrote a book on Relativity, published in 1940.

Around 1956, he became skeptical about the age difference and then completely denied the validity of the Special Theory of Relativity. This transformation is really surprising and difficult to understand. Since he was a well-known physicist and published in the prestigious journal *Nature*, people answered his articles and tried to show that he was wrong about the age difference, but in vain. The controversy went on for several years until he concluded that the Special Relativity is wrong and of course he advised physicists to open their eyes. The fights stopped because of the lassitude of physicists. Dingle continued to complain that it is not a scientific attitude to ignore him.

Shortly after the end of the Dingle controversy, a physicist, Sachs, opened a new controversy around 1973 and without denying the Relativity Theory, he claimed that there is no age difference; the traveler comes back with the same age as the observer on the earth. This time, the controversy was short: Sachs published his article in *Physics Today* and in a subsequent issue some reactions were published. There were also some articles between Sachs and other physicists.

2.7. The Deniers

The deniers of the Theory of Relativity (Special and General) are numerous, and one can find a list of some of them in *Wikipedia*. The critics of the Theory of Relativity are all very different, each critic choosing a different point for attack: the "errors" of Einstein, the illogical aspects of the Special Relativity, or new interpretations of some crucial experiments.

Here we shall mention an association called the National Philosophy Alliance, which does not exist today. Until 2013, it was active with publications and meetings, but an internal conflict put an end to its activities. In fact, it continues its activities under a new name "John Chappell National Philosophy Alliance". It seems that this new Alliance wants to be more respectable than its precedent, invoking essentially the right to think independently of the mainstream consensus.

If we recall this National Philosophy Alliance, the reason is because its members considered that the Twin paradox is a strong argument against the validity of the Special Relativity.

We quote from an Internet site that does not exist today:

> The NPA study concluded that, after 100 years of work on this famous problem in special relativity, the Twin Paradox continues to be unresolved.

And at that time (2013), it challenged the "mainstream" to propose a solution:

> We ask that a single spokesperson or group be selected by a top physics organization such as the American Institute of Physics (AIP) or the American Association for the Advancement of Science (AAAS) to give the currently accepted solution and respond to questions about that solution.

We did not discover in the actual publications of the new Alliance this challenge. It is likely that its members have left this provocative position.

2.8. The Article

Titolo:

Scientia

Editore:

Bologna: Zanichelli

Data:

1910–1988

Collezione:

Scientia (1907–1988)

Informazioni e crediti

Risorse esterne

Accesso all'indice del: Vol·10 (1911)

Ricerca articoli

Scheda bibliografica e localizzazioni della copia cartacea

Informazioni sul formato DjVu

Il DjVu fornisce funzionalità di zoom, rotazione della pagina e, per alcune opere, ricerca e selezione del testo·

- Scarica il plug-in
- Come usare il plug-in

URL: http://amshi

L' ÉVOLUTION DE L' ESPACE
ET DU TEMPS

L'attention des physiciens s'est trouvée récemment ramenée vers les notions fondamentales de l'espace et du temps que de nouveaux faits expérimentaux les obligent à remanier; rien ne peut mieux montrer l'origine empirique de ces notions que leur adaptation progressive, non terminée encore, aux données de plus en plus subtiles de l'expérience humaine.

Je voudrais montrer que la forme, insuffisamment analysée d'ordinaire, sous laquelle ces notions se présentaient jusqu'ici, était déterminée, conditionnée, par une synthèse particulière et provisoire du monde, par la théorie mécaniste. Notre espace et notre temps étaient ceux exigés par la mécanique rationnelle.

A la synthèse nouvelle, de plus en plus puissante, que représente la théorie électromagnétique des phénomènes physiques, correspondent un espace et un temps, un temps surtout, autres que ceux de la mécanique, et en faveur desquels nos moyens actuels d'investigation expérimentale viennent de se prononcer. Il est particulièrement remarquable que le perfectionnement croissant de nos méthodes de mesure, dont la précision a pu être poussée pour certaines au-delà du milliardième, nous oblige à continuer encore aujourd'hui l'adaptation aux faits des catégories les plus fondamentales de notre pensée. Il y a là, pour le philosophe, une occasion excellente de pénétrer la nature intime de ces catégories en les trouvant encore en voie d'évolution, en les voyant vivre et se transformer sous ses yeux.

Chapter 3

The Article of Langevin

We begin with Langevin's article of 1911, because first, it is logical to begin with the beginning and, secondly, it is an excellent presentation on the Relativity Theory, maybe better than the exposition in numerous textbooks. The article of Langevin presents the Relativity Theory in a qualitative way without any formulas. In the following chapter, we present the same notions but in a quantitative way with the help of mathematical formulation.

The Evolution of Space and Time

The attention of physicists was recently directed towards the fundamental notions of space and time, as new experimental facts oblige us to alter them; nothing can better demonstrate the empirical origin of these notions than their progressive adaptation, still not finished, with the increasing subtlety of human experiments.

I would like to show that the form, usually insufficiently analyzed, in which these notions arose, up to the present, was determined and conditioned by a particular and provisional synthesis of the world by the mechanistic theory. Our space and time were given as required by rational mechanics.

The new synthesis, which is increasingly powerful and which is represented by the electromagnetic theory of physical phenomena, corresponds to space and time (especially time) that are different from those of mechanics and that are favored by our current methods of experimental investigation. It is particularly remarkable that the increasing perfection of our measurement methods, whose accuracy has been pushed beyond a billionth, obliges us to continue the adaptation to the facts on the most fundamental categories of our thought. For the philosopher, there is an excellent opportunity to penetrate the intimate nature of these categories, while they are still in the process of evolution, to see them thriving and transforming before his eyes.

There exists neither space nor time *a priori*: to every moment, every level of perfection of our theories of the physical universe corresponds to a conception of space and time. Mechanics implies the old conception, electromagnetism requires a new one, and nothing permits us to declare that this is the definite one.

It is also difficult for our brain to get used to these new forms of thought: reflection is particularly delicate and can only be aided by the formation of an adequate language ability.

This is the task, to facilitate the evolution of mankind, for which today philosophers and physicists should work together.

All living beings have the ability to expand internally and spontaneously, which is even greater when they have adapted better to the milieu in which they arose. If, as a result of this expansion, an encounter takes place between individuals or species, there may occur mutual adaptation or, if accordance is impossible, there occurs a conflict resulting in the survival of the fittest, which generally assimilates the substance of the other and imposes on it a new form that life seems to have judged better.

The same is the case for our physical theories: some are particularly well established, and have succeeded brilliantly in the interpretation and grouping of a category of experimental facts, to which they impose a form and then they develop (spontaneously and in agreement with this form) this rhythm of their own, by taking the already known but scattered facts as substance of the building which they construct, and those facts to which they are directed to learn, and finally those facts already established as synthesis in the form of various theories that will be absorbed by a new one after coming in to conflict with them.

Just as the effort of living beings to grow is facilitated by the organic syntheses already realized in other beings that were absorbed, a new theory conserves and uses more or less completely the body of facts already established by the theories over which it has triumphed.

We are now witnessing a conflict of this kind between two conceptions of the universe that are of particular importance and beauty: on the one hand, the rational mechanics of Galileo and Newton and, on the other hand, the electromagnetic theory in the advanced form as it was given by Maxwell, Hertz, and Lorentz.

Rational mechanics was created for the interpretation of the phenomenon of visible motion and has succeeded admirably. All the scientific effort of the 18th century and much of the 19th century was devoted to extending this ability to explain all physical phenomena by applying these laws to the motion of various invisible material particles or fluids.

In this way, the doctrine known as mechanics developed from the fusion of rational mechanics and the atomistic hypotheses. The success was great in certain domains, such as the kinetic theory of fluids, and less in other domains such as elasticity and optics.

We should not forget that it was the atomistic conception that was often made responsible for the failure of mechanics; today, however, it is definitively established by indisputable experimental facts, and its association with the electromagnetic theory proved its remarkable fertility in the last 15 years. What really seems to be questionable is the application of mechanical laws to invisible motions, which were at first established for visible motions, and even for them they represent only a first approximation, albeit excellent.

The theory of electromagnetic phenomena as we have it today is certainly independent of the laws prescribed on the motion of matter by rational mechanics, although it appears to be involved in certain fundamental definitions: the best evidence of this independence is provided by the contradictions that currently exist between the two syntheses.

Electromagnetism is just as remarkably adapted to its original domain as rational mechanics was adapted to its domain, with its notions of a very special medium that transmits the actions step by step; the electric and magnetic fields characterize the state of this medium, with very particular forms of relations between the simultaneous variation of these fields in space and time; electromagnetism constitutes a discipline or, a way of thinking that is quite different, and quite separate from mechanics and is endowed with an immense force of expansion that has assimilated the immense domains of optics and radiant heat to which mechanics remains incapable, and every day it provokes new discoveries in this field. Electromagnetism has conquered much of physics, has invaded chemistry, and has grouped a large number of facts that were hitherto formless and disconnected.

Of our two opposing theories, the first possesses the merit of a long-standing history and the authority of having seen the verification of its laws by the most distant star and the molecules of the most tenuous gas; the second, which is younger and more alive, is much better adapted to the entirety of physics and possesses an inner force of growth that the other seems to lack.

Maxwell thought that it's possible to reconcile the two theories and to show that electromagnetic phenomena are susceptible to mechanistic interpretations, but his demonstration, published elsewhere for the particular case of phenomena presented by closed currents, proves only that the two syntheses share common characteristics, such as the common property of leaving certain integrals stationary, but they may remain irreconcilable in other characteristics.

These differing characteristics have recently been demonstrated by new experimental facts and by the negative result of all experiments, some of extraordinary delicacy, which were performed to try to demonstrate the total uniform translational motion of a material system by experiments within this system and to show the absolute translational motion.

We already know, and rational mechanics is perfectly consistent with this fact, that mechanical experiments on visible motions, performed within a material system, do not permit to demonstrate the uniform translational motion of the entire system, but on the contrary, for rotational motion it can be achieved by using the Foucault pendulum or the gyroscope. In other words, according to the mechanical point of view, the collective uniform translation has no absolute sense; rotation, on the contrary, has one.

But within a material system, other experiments can be tried which involve electromagnetic or optical phenomena. Electromagnetic theory involves a medium for its explanation, the ether, which transmits the electrical and magnetic actions and in which electromagnetic disturbances (light, in particular) propagate with a determined velocity.

It was hoped that if a material system moves with a uniform translation with respect to this medium, electromagnetic or optical experiments inside of the system permit the demonstration of this translation.

Since the earth, in its annual motion, possesses a translation velocity which is constantly varying by up to 60 kilometers per second for the relative velocity corresponding to two diametrically opposed positions of the globe in the orbit, it was hoped that at least at certain times of the year, the observers on the earth and their equipment will move in relation to the ether with a velocity of this order and might be able to demonstrate their motion.

This could be expected because, by combining the fundamental equations of electromagnetism, which are believed to be accurate for stationary observers in the ether, with the ordinary notions of space and time as required by rational mechanics, these equations would change their form to the observers moving in the ether, and the differences in velocities like that of the earth in its orbit should be visible in certain experiments of extraordinary delicacy.

But the result was found consistently negative, and independent of any interpretation, we can state as an experimental fact the content of the following principle, namely that of relativity:

> If different groups of observers are in uniform translation against each other (such as observers attached to the Earth for different positions of the latter in its orbit), all mechanical and physical phenomena follow the same laws for all these groups of observers. None of them, by experiments inside of the material system to which they are attached, can demonstrate the uniform translation of the whole system.

From the electromagnetic point of view, we can still say that the fundamental equations, in their usual form, are valid for all these groups of observers at the same time, and that everything happens for each one as if it were stationary relative to the ether.

So it's an experimental fact that the equations representing physical quantities with which we translate the laws of the outside world must have exactly the same form for different groups of observers for various systems of reference in uniform translation relative to each other.

This requires, in the language of mathematics, that these equations allow a group of transformations corresponding to a change of the reference system to another moving relative to it. The equations of physics must be preserved for all transformations of this group. In such a transformation, when one moves from one reference system to another, measures of various magnitudes, especially those that are related to space and time, are changed in a manner that corresponds to the structure of these notions.

Now, the equations of rational mechanics actually allow a group of transformations corresponding to the change of the reference system, and the part of that group that is related to measurements of space and time is in accordance with the usual form of these notions.

The great merit of H. A. Lorentz is to have shown that the fundamental equations of electromagnetism also allow a group of transformations that in turn allow them to take the same form when we pass from one reference system to another; *this group deeply differs from the preceding one concerning the transformations of space and time.*

We must choose as follows: if we want to maintain an absolute value for the equations of rational mechanics and of space and time corresponding to them, we must consider as false those of electromagnetism, and reject the admirable synthesis that I mentioned above, and introduce for example an optical emission theory with all the difficulties it entails and which was dismissed over 50 years ago. However, if we want to keep electromagnetism, we must adapt our minds to the new concepts that are required for space and time and consider rational mechanics as having no more value than the value of a first approximation, but it is largely sufficient when it comes to motions whose velocity does not exceed a few thousand kilometers per second. Electromagnetism, or the laws of mechanics admitting the same transformation group as electromagnetism, alone permits to go further and take the important place of rational mechanics.

To highlight better the contrast between the two syntheses, it is useful to merge, as proposed by Minkowski, the two notions of space and time in the more general notion of the world.

The world is an ensemble of all events: an event consists in the fact that it happens, or there is something in a certain place at a certain moment. If a reference system is given, that is to say a system of axes connected with a certain group of observers, any event is determined in terms of its position in space and time by four coordinates measured in this reference system, three for space and one for time.

If two events measured in a certain reference system are given, they generally differ in both space and time, and occur in different places at different times. To a couple of events, these corresponds a spatial distance (the points where the two events are happening) and a temporal interval.

We can define time by all the events that follow one another at one point, for example, in the same portion of matter in relation to a reference system, and define space by all the simultaneous events. This definition of space corresponds, in effect, to that the shape of a moving body is defined by the set of simultaneous positions of the various portions of matter it contains, its various material points, or by all events posed by the simultaneous presence of these different material points. If we agree to call, with Minkowski, the *world line* of a portion of matter that may be in motion relative to the reference system, all events that occur in that portion of matter, then the shape of a body at a given moment is determined by the set of simultaneous positions on the world lines of various material points which constitute this body.

The notion of simultaneity of events happening at different places is fundamental for the very definition of space when a body in motion is concerned, and this is generally the case.

In the ordinary conception of time one attributes to simultaneity an absolute sense, supposedly independent of the reference system; it is necessary that we analyze more closely the content of this generally tacit hypothesis.

Why do we usually not admit that two events that are simultaneous for a certain group of observers may not be simultaneous for another group moving relative to the first, or, equivalently, why don't we admit that a change of reference system permits to reverse the order of temporal succession of two events?

This is obviously a consequence of our implicit admission that if two successive events occur in a certain order for a given reference system, the one that occurred first would be able to intervene as a cause and change the conditions under which the second occurs, regardless of their spatial separation.

Under these circumstances, it is absurd to suppose that for other observers for another reference system, the second event, the effect, may occur before its cause.

The absolute nature, usually admitted in the notion of simultaneity, is a consequence of the implicit hypothesis that causality can propagate with infinite velocity, and the hypothesis that an event can occur simultaneously as the cause at any distance.

This hypothesis is consistent with the mechanistic conception since it is required by the concept of a perfectly rigid body of rational mechanics; for example, an inextensible bell cord that is interposed between the two points where the events occur would instantly signal the occurrence of the first event to the point where the latter will occur, and would consequently allow to take the first into account, to utilize it as the cause under the conditions which determine the second. So there is mutual adaptation of the rational mechanics and ordinary conceptions of space and time in which the simultaneity of two distant events in space has an absolute sense.

We are therefore, not surprised to find that in the transformation group which preserves the equations of mechanics, *the time interval of two events is conserved, and it is measured in the same way by all groups of observers, regardless of their relative motions.*

It is different in the case of the spatial distance: it's a simple fact and contained in the usual notions that the spatial distance of two events generally has no absolute sense and depends on the reference system that is used.

A concrete example will show how the spatial distance of the same two events may be different for different groups of observers in relative motion to each other. Imagine a hole in the floor of a car in motion relative to the ground, and then one drops two objects in succession: the two events that constitute the exits of the two objects through the hole occur at the same point for observers related to the car, but at different points for observers related to the ground. The spatial distance of these two events is zero for the first observers, but for the other it is equal to the product of the velocity of the car with the time interval between the fall of the two objects.

It is only if the two events are simultaneous that their distance in space has an absolute sense, so that they don't vary with the reference system. It follows immediately that the dimensions of an object, for example the length of a ruler, also have an absolute sense and are the same for the observers at rest or in motion relative to this object: we have noticed that for any observation, the length of a ruler is the distance between two simultaneous positions of the ends of the ruler, that is to say, the spatial distance of two simultaneous events, two simultaneous occurrences of both ends of the ruler. We have seen that simultaneity, as well as the spatial distance of two simultaneous events, has an absolute sense in the usual conception of time and space.

Given any two successive events, i.e., two events separated in time, we can always find a reference system in which these two events coincide in space, with observers to whom these two events happen at one point. It will indeed be sufficient to give these observers, compared to the original reference system, a motion so that they attend the first event and next they attend the second event, so that for them the two events occur at the same point close to them; it suffices to give the observers a velocity equal to the ratio of the spatial distance to the time interval between both events in the original reference system, and this is always possible if the time interval is not zero, i.e., if the two events are not simultaneous.

What can be achieved for space, namely the coincidence of two events in space by a suitable choice of reference system, cannot be achieved for time, since the time interval of two events has an absolute sense; it is measured in the same way in all reference systems.

There is an asymmetry between space and time as they are habitually given that disappears in the new concepts: the interval in time, as well as the distance in space, will become variable with the reference system, that is, with motion of the observers.

In the new concepts, only one case remains and must remain where the change of reference system is ineffective: it is where the two events coincide in both space and time: this double coincidence must indeed have an absolute sense, since it is the encounter of the two events, and this encounter may produce a phenomenon, a new event, which necessarily has an absolute sense. Remember the previous example; if the two objects coming out of the car by the same hole leave simultaneously, i.e., if their endpoints coincide in both space and time, it may result in a shock and the breaking of the objects, and this shock phenomenon has an absolute sense, so that in any conception of the universe, electromagnetic or mechanical, the coincidence in both space and time, if it exists for a group of observers, cannot be denied by another group, regardless of their motion relative to the first. For those who see the car passing by, as for those who are there, the two objects will break each other because they come together at the same point.

Except for this very special case, it is easy to see that the electromagnetic conception requires a major overhaul of the notion of the world. The equations of electromagnetism imply, in their usual form, that an electromagnetic disturbance, e.g., a light wave, propagates in vacuum with the same velocity in all directions, equal to 3,000,000 km/s.

The newly established experimental facts have shown that if these equations are exact for a group of observers, they should also be exact for all others, regardless of their motion relative to the first, resulting in the paradoxical fact that a light disturbance must be spread with the same velocity for different groups of observers in motion relative to each other. A first group of observers sees a light wave propagating in a certain direction with a velocity of 3,000,000 km/s and sees another group of observers following this wave with an arbitrary velocity; however, for this second group, the light wave will move relative to them with the same velocity of 3,000,000 km/s.

Einstein first showed how this necessary consequence of the electromagnetic theory is sufficient to determine the characteristics of space and time required by the new conception of the world. It is conceivable, according to the above, that the velocity of light must play an essential role in the new formulations: it is the only velocity that is preserved when passing from one reference system to another and plays in the electromagnetic world the role which is played by the infinite velocity in the mechanical world. This will be clear from the results that follow.

For any pair of events, changing the reference system changes both the distance in space and time interval, but in view of the importance of these changes, we are led to classify couples of events into two major categories for which time and space play symmetric roles.

The first category consists of pairs of events for which their spatial distance is greater than the path traveled by light during the time interval; that is to say, when the emission of light signals accompanies the production of two events, each will take place *before* the passage of the signal from the other. Such a relationship has an absolute sense; that is to say, it is valid for all reference systems, if it is valid for one of them.

The transformation equations required by the electromagnetic theory show that in this case the order of succession of two events in time has no absolute sense. If, for a first reference system, the two events follow each other in a certain order, this order will be reversed for observers moving with respect to the first with a velocity less than that of light, that is to say, at a physically attainable velocity.

It is evidently impossible for two events whose order of succession may be reversed and are united by a relationship of cause and effect, because if such a relationship would exist between our two events, some observers would see the cause after the effect, which is absurd.

However, since the spatial distance of our two events is greater than the path traveled by light during their time interval, the first cannot be involved in the occurrence of the other and the second cannot be informed of the first, except if the causal connection could propagate with a velocity exceeding that of light. We must therefore, according to the above, eliminate this possibility: causality, whatever its nature, must not propagate with a velocity exceeding that of light; there should be no messenger or signal which can travel at more than 3,000,000 km/s.

We must therefore admit that an event cannot act instantly as a remote cause and that its impact can only be felt immediately on the position or the point where it took place, and subsequently at increasing distance, increasing with, at most, the velocity of light. In this view, the new concept plays the same role as that played by the infinite velocity in the old concept, and represents the velocity limit at which causality can propagate.

Hence, we see that the current antagonism between mechanics and electromagnetism only manifests in a new form, the opposition between two conceptions which followed one another during the development of electrical theories: that of instantaneous action at a distance consistent with mechanics, and transmission through a medium by direct action as introduced by Faraday. This old opposition occurs nowadays even for the most fundamental concepts.

From the foregoing, various consequences emerge: firstly, it is impossible for a portion of matter to move in relation to another with a velocity exceeding that of light. This paradoxical result is contained in the formulas that led to the new kinematics of velocities: the composition of any number of velocities below the velocity of light always gives a lower velocity than that of light. Similarly to the ordinary conception, the composition of any number of finite velocities still gives a finite velocity.

We can then say that no action at a distance, such as gravitation, can propagate faster than light and we know that this condition is not refuted by the astronomical results established currently.

Finally, it is necessary to abandon the perfectly rigid body of mechanics in which we could find a way to signal immediately at a distance to establish a causal connection to propagating faster than light. Nothing of what we know about real rigid bodies is opposed to this fact that any action or any wave must propagate at a lower velocity than light; elastic waves in the most rigid solids, in fact, propagate with a much lower velocity. The important thing is that we must reject the conception of a perfectly rigid body, a body that could be put into motion at all its points simultaneously.

We can summarize the above reasoning as follows: if there were a signal that could propagate with a velocity exceeding that of light, observers could be found for whom this signal would have arrived before it departed and for whom the causal connection allowed by the signal would be reversed: we could telegraph in the past, as it was said by Einstein, and we consider that this would be absurd.

The two events of the pair in question, with no definite order of succession in time, are thus necessarily without possible mutual influence, so they are truly independent events. Evidently, having no causal connection between them, they cannot follow each other in the same portion of matter, and cannot belong to the same world line in the life of a single being. This impossibility is in accordance with the fact that, to be successively the seat of these two events, this portion of matter should move with a velocity exceeding that of light.

Both events cannot be brought to coincide in space for any choice of reference system, but they can be brought to coincide in time: since their sequence can be reversed, there are reference systems for which the two events are simultaneous.

We can call *pairs in space* those pairs of events that have just been considered and to which the order of succession in time has no absolute sense, but are spatially distant in an absolute way.

It is worth noting that, while the spatial distance of two events cannot be canceled, it reaches a minimum precisely for reference systems in which the two events are simultaneous.

Hence the following statement:

The spatial distance of two events that are simultaneous for a certain group of observers is shorter for them than for all other observers in arbitrary motion relative to them.

This statement contains, as a particular case, what is called the Lorentz contraction, that is to say, the fact that the same ruler considered by different groups of observers, some resting, others in motion relative to it, is shorter for those who see it passing by than for those who are attached to it. We have already seen that the length of a ruler for observers who see it passing by is defined by the distance in space of two simultaneous positions (for those observers) on both ends of the ruler. According to the preceding, this distance will be shorter for these observers than for all others, especially those attached to the ruler.

We also easily understand how the Lorentz contraction can be reciprocal, that is to say, how two rulers that are equal when at rest appear mutually shortened when they slide against one another, and observers attached to one of the rulers will see the other one shorter than its own. This reciprocity holds, because observers associated with the two rulers in motion relative to each other do not define simultaneity the same way.

We will find for the pairs of events of the second category, properties exactly correlative to the preceding one by permutation of space and time. These pairs, which I will call *pairs in time*, are defined by the following condition, which has an absolute sense: the spatial distance of two events is less than the path traveled by light during their time interval; in other words, the second event occurs *after* the passage of the light signal whose emission coincides in space and time with the first. This introduces, from the point of view of time, an asymmetry between the two events, as the first occurs before the passage of the light signal whose emission coincides in space and time with the second event, while the second occurs after the passage of the light signal which accompanies the first one. A causal connection may exist at least by light between the two events, so the second has been informed of the first, and this requires that the order of succession has an absolute sense, and cannot be reversed by any change of the reference system. We immediately see that such a reversal would require a velocity exceeding that of light for the second reference system with respect to the first.

Two events between which there exists a real possibility of influence, if they cannot be made to coincide in time, they can always be made to coincide in space by a suitable choice of the reference system. In particular, if the two events belong to the same world line, following each other in an absolute order in the life of a portion of matter, they will coincide in space for observers attached to that portion of matter.

In correlation with what was happening earlier, if the time interval of two events cannot be canceled, it passes by a minimum precisely for the reference system in which the two events coincide in space.

Hence the statement:

The time interval between two events that coincide in space, which follow each other at the same point for a certain reference system, is less for it than that for any other system in arbitrary uniform translation relative to the first.

In all the above, the employed reference systems are supposed to possess uniform translational motion: for only such systems, observers associated with them cannot experimentally detect their collective motion, and only for such systems the equations of physics must hold their shape when switching from one to another. For such systems it is thus as if they were stationary relative to the ether: a uniform translation in the ether has no experimental sense.

But it should not be concluded because of this, as has been sometimes done prematurely, that the concept of ether must be abandoned and that the ether is non-existent and inaccessible to experiment. Only a uniform velocity relative to it cannot be detected, but any change of velocity or any acceleration has an absolute sense. In particular, it is a fundamental point in the electromagnetic theory that any change of velocity or any acceleration of an electrified center is accompanied by the emission of a wave that propagates in the medium with the velocity of light, and the existence of this wave has an absolute sense; and conversely any electromagnetic wave, for example light, has its origins in the change of the velocity of the electrified center.

> We therefore have a hold on the ether through accelerations, and acceleration has an absolute sense as determining the production of waves from matter that has undergone a change in velocity, and the ether manifests its reality as the vehicle, as the carrier of energy transported by these waves.

Up to this point, Langevin has considered only reference systems in uniform translational motion. One can say that this corresponds to the Theory of Special Relativity since in this theory there is no change in the velocities. In the following, Langevin includes in the translational motion a variation of the velocity, the acceleration. In fact, this is more in accordance with the Theory of General Relativity, which considers systems with acceleration, as we shall see in Chapter 4.

It is worth noting that Langevin explains very well the reciprocity of the observations made by people in different reference systems. So the measurement of the ruler of one system may be made by other reference systems with different results. The same thing for measurements of time: the time interval between two events is seen or measured differently in different systems, but there is no encounter between people of different systems. One considers here only measurements about events which occur in a particular system by people in other systems.

One notes that Langevin proposes not to leave the concept of the ether and hopes that an experiment will be found in order to detect it. As is well known, in the Theory of Relativity there is no place for the ether.

Langevin goes on with the introduction of acceleration and reaches new conclusions: now there will be an encounter between people of different systems.

> The theory provides the opportunity to demonstrate, by electromagnetic or optical experiments, any acceleration of the collective motion of a material system by means of experiments inside this system, if only on finding the wave emission by electrified

bodies related to the system, which are immobile relative to it. We also know that if the acceleration of the collective motion is communicated to the system by external actions that are exerted, contrary to what happens with gravity, only on parts of the system, we have many other means to demonstrate it, for example, strains within the system through which the acceleration is transmitted from portions of the system that suffer external actions to other portions that do not undergo them.

In a uniform gravitational field, where each portion of the system would suffer direct external action, which would communicate the overall acceleration, as in the projectile of Jules Verne, similar reactions do not occur, but as I said above, the possibility of using electromagnetic or optical experiments to detect the change in velocity of the collective motion would remain: the laws of electromagnetism with respect to axes attached to this material system are not the same as those with respect to axes in collective uniform motion of translation.

We will see the appearance of this absolute character of acceleration in another form.

Consider a portion of matter in arbitrary motion and the sequence of events that constitute the life of this portion of matter, its world line.

For two of these events that are sufficiently close to one another, observers in uniform motion who attend these two events successively can be regarded as related to this portion of matter, the velocity change of the latter being imperceptible in the interval between the two events. For these observers, the time interval between the two events constitutes an element which we call the *proper time* of the portion of matter; it will be shorter than that for any other group of observers associated with a reference system in arbitrary uniform motion.

If we now take any two events in the life of our portion of matter, then their time interval measured by observers in non-uniform motion who will constantly monitor the portion of matter will be, by integrating the previous result, shorter than that for the reference system in uniform motion.

In particular, in this reference system the two events considered may be taking place at the same point, in relation to which a portion of matter has traveled a closed cycle and has come back to its starting point thanks to its non-uniform motion. *And we can say that for observers related to that portion of matter, the time period elapsed between the departure and return, i.e., the proper time of the portion of matter, will be shorter than that for observers who would have stayed connected to the reference system in uniform motion.* That portion of matter would have aged less between its departure and its return than if it had not been accelerating, i.e., if it had remained stationary relative to a reference system in uniform translation.

We can still say that it is sufficient to be agitated or to undergo accelerations, to age more slowly, and we'll see in a moment how much you can expect to get in this way.

Giving concrete examples: imagine a laboratory attached to the earth, whose motion can be considered as uniform translation, and in this laboratory there are two perfectly identical samples of radium. What we know about the spontaneous evolution of radioactive materials allows us to say that if these samples are kept in the laboratory, they both will lose their activity the same way over time and their activities will remain continuously equal. But send one of these samples with a sufficiently high velocity and then bring it back to the laboratory; this requires that at least at certain times this sample has undergone accelerations. We can say that on return, its proper time between departure and return is less than the time interval measured between these events by observers attached to the laboratory, so that it has evolved less than the other sample and therefore it will be more active than the latter; it would have aged less, having been more agitated. The calculation shows that for a difference of one ten-thousandth between the changes in activity of the two samples, it will be necessary to maintain (during separation) the velocity of the traveling sample at approximately 4000 km/s.

Before we give another example, let us present our result in a different light. Suppose that two pieces of matter meet for the first time, separate, and meet again. We can say that observers attached to the portions during the separation will not evaluate that duration in the same way, as some have not aged as much as the others. It follows, from the foregoing, that those have aged the least for whom the motion during separation was most distant from uniform, and who have suffered the greatest accelerations.

This remark provides a means, for any among us who wants to devote two years of his life, to find out what the earth will be like in 200 years, and to explore the future of the earth, by making in his life a jump ahead that will last two centuries for the earth and for him it will last two years, but without hope of return, without a possibility of coming to inform us of the result of his voyage, since any attempt of the kind could only transport him increasingly further.

For this, it is sufficient that our traveler consents to be locked in a projectile that would be launched from the earth with a velocity sufficiently close to but lower than that of light, which is physically possible, while arranging an encounter with, for example, a star that happens after one year of the traveler's life, and which sends him back to the earth with the same velocity. Returned to the earth he has aged two years, then he leaves his ark and finds our world 200 years older, if his velocity remained in the range of only one twenty-thousandth less than the velocity of light. The most established experimental facts of physics allow us to assert that this would actually be so.

It is amusing to realize that our explorer and the earth would see if they could, by light signals or wireless telegraphy, mutually stay in constant communication during their separation, and thus understand how asymmetry between two measures of the duration of separation is possible.

As they move away from each other with a velocity close to that of light, each seems to flee before the electromagnetic or optical signals that were sent to the other, so that there will be a very long time to receive signals that were emitted during a given time. The calculation shows that each of them will see the other live 200 times slower than usual. During the year of this distant motion, the explorer will receive the news from the earth of the first two days after his departure, and during this year he will see the earth make the motions of two days. Moreover, for the same reason, the radiation he receives from the earth during this time has wavelengths that are 200 times greater due to the Doppler principle. What seems to him as light radiation by which he can see the earth has been emitted as extreme ultraviolet radiation, maybe close to Roentgen rays. And if one wants to maintain communication between them by Hertzian signals or by wireless telegraphy, the explorer who brought with him a receiving apparatus with a certain antenna length and the transmission devices used by the earth during these two days of departure must.

During the return, the conditions are reversed: for each of them sees the life of the other as remarkably fast, 200 times faster than usual, and during the year for him to return, the explorer sees the earth perform the actions of two centuries: so it can be understood how he can find the Earth aged by 200 years. During this period he will also see waves that are bright for him, but they will be emitted as far-infrared rays of around 100 microns in wavelength as Rubens and Wood have recently discovered in the emission spectrum of a Welsbach mantle. For him to continue to receive radio signals of the Earth, it shall, after the first two days and for two centuries thereafter, use a transmitting antenna 200 times longer than that of the traveler, 40,000 times longer than that used during the first two days.

To understand the asymmetry, it is worth noting that the earth will spend two centuries to receive the signals sent by the explorer during his motion away from it which lasted a year: he will live during this time in his ark a life 200 times slower, while he will make the gestures of a year. In the two centuries during which the earth sees the observer moving away, it must, to receive the Hertzian signals emitted by him, use an antenna

200 times longer than that of the observer. At the end of two centuries, the news of the encounter of the projectile with the star will come to earth, which marks the beginning of the return voyage. The arrival of the traveler will occur two days afterwards during which the earth will see him living 200 times more quickly than usual, and it will see him achieving the gestures of another year to find him only two years aged at the return. During the last two days, to receive news of him, the earth will have to use a reception antenna 200 times shorter than the antenna of the traveler.

Thus the asymmetry — which occurred because only the traveler, in the middle of his journey, has undergone an acceleration that changes the direction of his velocity, and which brings him back to the starting point on earth — results in the fact that the traveler sees the earth moving away and approaching during times that are equal to one year, while the earth, which is informed of the acceleration only by the arrival of light waves, sees the traveler moving away during two centuries and returning during two days, i.e., during a time 40,000 times shorter.

Now, if we look for the conditions under which a similar program could be realized, one encounters, of course, considerable physical difficulties.

The theory permits to calculate the work that the earth should spend to launch the projectile and to transmit the kinetic energy corresponding to its enormous velocity. Assuming the mass of the projectile is only equal to one ton, it can easily be calculated if one wants to spend only one year to launch it, for example by turning it at the end of a sling before releasing it, then it should function non-stop for that year by 400 billion horsepower, and to produce them it must burn at least 1000 km^3 of coal.

These difficulties at the beginning would be followed by difficulties no less large at the time of the reflection or stop. We should, first of all, for reflection, find a system capable of storing the enormous kinetic energy of the projectile, and then restore it to return it in the opposite direction with the same velocity. To stop, one should gradually dissipate this energy without resulting, at any moment, in an acceleration or temperature elevation harmful to the projectile, while the amount of heat equal to its kinetic energy is sufficient to raise it to a temperature of 10^{16} degrees or more.

We also have every reason to think that if a projectile is coming toward the earth with such a velocity, it wouldn't even notice its passage and would stop only at a certain depth in the soil without leaving any holes in the same area of the surface where this would have happened. It would hardly produce on its trajectory through the atmosphere a slight increase in electrical conductivity of the air. We know, indeed, by the example of α-particles of radium, i.e., the material atoms of helium with a velocity of just $20{,}000\,\mathrm{km}$ per second, that theory can follow a perfectly straight trajectory and pass through other atoms without leaving any trace of their passage other than increased conductivity, and our projectile has, per unit mass, a kinetic energy $1{,}000{,}000$ times greater than α-particles. It would be an extraordinarily penetrating radiation. If we want to avoid these difficulties, we have to find a way to slow its motion gradually as it approaches the earth. It does not seem to be possible either to use the principle of the rocket that my friend Perrin proposes for interplanetary travel.

I developed these speculations only to show, by a striking example, what consequences, far away from the usual concepts, the new form of the concept of space and time leads to. It must be remembered that this development is based on perfectly correct determinations as required by indisputable experimental facts, which our ancestors were not aware of when they constituted in their experience in relation to synthesis mechanics, the categories of space and time that we have inherited from them. It's up to us to extend their work by pursuing with greater detail, with the means at our disposal, the adaptation of thought to the facts.

It is not only in the field of space and time that the redesign of the most fundamental concepts of the mechanistic synthesis is required. The mass, which was measured by inertia, an essential attribute of matter, was regarded as an essentially invariable element, characterizing a given portion of matter. This notion now vanishes and merges with that of energy: the mass of a piece of matter varies with the internal energy of it, rises and falls with it. A piece of matter which radiates loses its inertia by an amount proportional to the energy radiated. It is the energy which is inert; matter can resist the change in velocity only in proportion to the energy it contains.

The concept of energy itself loses its absolute sense: its measurement varies with the reference system that the phenomena are related to, and physicists are currently, in the expression of the laws of the universe, looking for the real elements that possess an absolute sense, i.e., the elements that remain invariant when changing from one reference system to another, and which will play the same role in the electromagnetic conception of the world that was played by time, mass and energy in the mechanistic synthesis.

Paul Langevin
Collège de France, Paris

In order to have a better understanding of the signal exchange as described by Langevin, we shall use the space–time diagram in the reference system of the stay on the earth. The x-axis is the distance of the traveler and what is in general the y-axis is the axis "ct", the time multiplied by the light velocity. This explains why the stay at home receives the knowledge of the arrival on the star only two years after the departure and why for the stay at home on the earth, the return is observed only after two years (Fig. 3.1).

Now we shall make some important remarks about this article.

The first concerns the name "Twin paradox". These words do not appear in the article. It is clear that for Langevin it is not a paradox but only a consequence of the theory. So there is no need for the word "paradox".

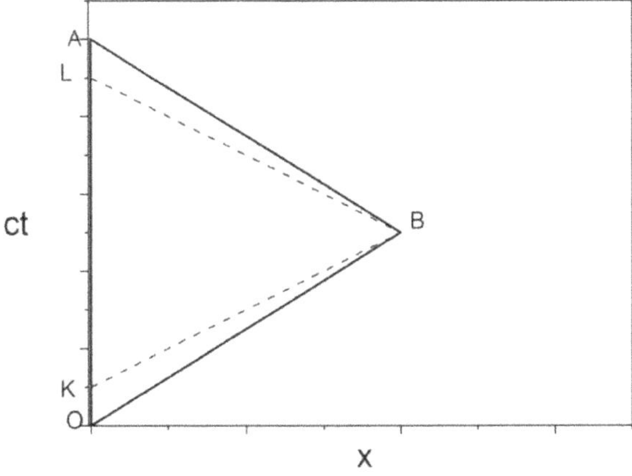

Fig. 3.1. Space–time diagram of the Langevin traveler. OA is the trajectory of the earth and the lines OB and BA are that of the traveler. BK is the last light signal received by the earth when the traveler reaches the star and BL is the light signal sent by the traveler when he turns out at the star. For the earth inhabitants the return time of the traveler is only the distance LA. The time OB is one year like the distance BA but the distance OA is 200 years.

Secondly, the article is only qualitative without mathematical formulas which would permit to see exactly how the Relativity Theory predicts the difference in the ages of the two observers: the stay on the earth and the traveler. This is the reason why (in particular after 1950) there were so many articles proposing the explanation. If Langevin had presented the exact formulas, the whole "Twin paradox" would not have appeared. And when effectively his solution was published, it was in a not well known French journal.

Thirdly, Langevin attributes the dissymmetry and the age difference to the acceleration. As he writes:

> We can still say that it is sufficient to be agitated or to undergo accelerations, to age more slowly, and we'll see in a moment how much you can expect to get in this way.

The difference in age given by Langevin seems exaggerated. He states that "his velocity remained in the range of only one twenty-thousandth less than the velocity of light." It is clear that this velocity of the traveler is relative to the earth. In other words, if the stay on earth is 200 years older at the return of the traveler, this means that for him the travel to the star took about 100 years. Since the velocity is very close to the light velocity, we conclude that for him, the star is about 100 light years from the earth. It seems very difficult to understand how the traveler covers this distance in one year of his proper time. Nevertheless, the result of Langevin is correct as we shall see later (see Chapters 4 and 5): the stay at home is older than the traveler at their encounter.

Finally, Langevin is very explicit:

Returned to Earth he has aged two years, then he leaves his ark.

The last words are very important, the traveler stops his engine and leaves it. Consequently the two observers are in the same frame. We shall see later that a version of the paradox is proposed where there is no encounter and it is clear that this version is not that of Langevin.

The essential point of the article is that Langevin attributes the age difference to the acceleration of the traveler. Even if in a large part of the travel his velocity is constant, and even if one does take into account the acceleration at departure, it is impossible to ignore that, upon reaching his goal, he reverses his direction and this means the application of an acceleration. Sometimes, it was claimed that, if the time of reversal (the time of the application of the acceleration to change the direction of his velocity) is very short, this does not have an impact on the problem and one can ignore the acceleration. Later, we shall see that this is not correct, and even if the traveler bounces on the star to reverse his velocity, the acceleration has an important role in the problem.

PART II

PHYSICS

Chapter 4

The Clock or The Twin Paradox

In this chapter, we give a qualitative interpretation of the Clock paradox, also known as the Twin paradox. The word "twin" is used to give a more figurative aspect to the problem. However, we shall continue to call this phenomenon "the Clock paradox", although there is no paradox at all. This interpretation is based on the Special Theory of Relativity and also the General Theory of Relativity. This chapter is divided into three parts: first, a recall of the basic notions of Special Relativity, secondly, a basic discussion of the paradox based on Special Relativity, and finally, a recall of some notions of General Relativity to give a more complete picture.

Special Relativity

4.1. Inertial Frames

The world of Special Relativity is a world of frames moving one relative to others with constant velocity. It is simple to suppose that their motions are parallel to a particular direction that we call the "x" direction. A frame may be seen as a very large box in which all the objects are at fixed distances one relative to the others. In each frame, there are present the following:

(1) Clocks placed in each point of the frame so that it is possible to read their indications in all places. All the clocks are synchronized, i.e., they all indicate the same time.

(2) A reference system made of three perpendicular axes, x, y, and z. But, as was mentioned above, only one axis will be considered,

47

namely the x-axis. With the help of sticks, it is possible to measure the position of each point along the x-direction.

(3) Observers who can measure the position of a point and the duration of a process.

An inertial frame moves at constant velocity with respect to the other frames, but we need to be more precise about this sentence. In a closed frame, without window to see what is happening outside, the observer cannot decide whether he is moving (always relative to other frames) or at rest. He does not have any means to decide. Now, if he opens the window and sees the other moving frames, he has two possibilities: either to say that he is at rest and the other frames are moving (relative to him) or he can say that he is moving relative to the other frames. The basic assumption is that an observer will choose the first possibility.

The important relativity principle can be expressed as follows: all the observers in all the inertial frames agree that the laws of physics are the same in all the inertial frames. One consequence of this principle is that the velocity of light is the same in all the inertial frames and is the largest velocity matter may reach.

In the world, there happen what we call events. Something occurs at a definite place given by the x coordinate and at a given time t in a particular reference frame. This implies that the observer in his inertial frame has positioned a point as point "zero" on his x-axis and has chosen time "zero" for all the clocks. In fact, the observer can measure distances between points and durations of time.

It is supposed that an observer in an inertial frame can also measure the position and time of an event that occurs in a different inertial frame. We do not ask how this is possible, but one can imagine using exchange of light signals to make such measurements. The novelty of Special Relativity is that, contrary to our sensible experience, the time of an event measured by an observer in his own frame is not the same as that measured by another observer in another frame. There is no universal time that is the same for all the inertial frames.

We extend the principle of relativity to the case when we consider different frames. It is found that an observer who makes measurements of physical phenomena occuring in his own frame and another observer in another frame obey the same physical laws in the two frames. But this implies that the time is not the same in the two frames.

4.2. The Lorentz Transformation

In order to gain more information about the different ways of measuring the position and time of an event as measured by an observer in his own frame or another observer in another frame, we shall use the Lorentz transformation.

Consider two reference systems S and S' such that the x-axis of S is parallel to the x'-axis of S'. The time is measured in S as t and in S' as t'. The two systems move relative to each other with the velocity V parallel to x and x':

S' moves relative to S with the velocity V and S moves relative to S' with the velocity $-V$.

If an event takes place in S at (x, t), what are the values of x' and t' of this event measured by the observer in S'? The answer is obtained by the Lorentz transformation that can be written as

$$x' = \gamma(x - Vt), \tag{4.1}$$

$$t' = \gamma(t - Vx/c^2), \tag{4.2}$$

where c is the light velocity, $\beta = V/c$ and $\gamma = \dfrac{1}{\sqrt{(1-V^2/c^2)}} = (1 - \beta^2)^{-0.5}$.

In equations (4.1) and (4.2), it is implied that in the two systems S and S' their origins O and O' coincide (such that $x = x' = 0$) when the clocks in S and S' indicate the same time taken to be equal to 0. It is also possible to choose x and x' different from zero for $t = t'$ and a constant must be added to the right-hand side of (4.1).

From equations (4.1) and (4.2), one can obtain the values of x and t corresponding to an event (x', t') taking place in S'

$$x = \gamma(x' + Vt'), \tag{4.3}$$

$$t = \gamma(t' + Vx'/c^2). \tag{4.4}$$

4.2.1. *Application of the Lorentz transformation*

Now, one applies the Lorentz transformation to some cases. First, one considers a process composed of two events in the frame S'. One event, which is the beginning of the process, takes place in S' at $(x' = 0,\ t' = 0)$, and the other (which is the end of the process) takes place at $(x' = 0,\ t' = \tau_1)$, i.e., at the same place. The total time of this process in S' is τ_1. What is the time τ_2 of the same process measured in S? From (4.3) and (4.4), one obtains, for the end of the process as measured in S, $x = \gamma V \tau_1$ and $\tau_2 = \gamma \tau_1$. In other words, the time measured in S is longer than that measured in S', since $\gamma > 1$. In the literature, this is mentioned as the "time dilation". In other words, one wants to mention that for the S observer, the process that takes place in S' is longer in time; it is likely that for him, the clock of S' goes slower than his own clock.

We define the proper time of a frame by the time measured by the clocks of this frame. From what we saw above, the observer in S (who measures the time of the process that occurs in S') may calculate the proper time τ_1 of the frame S' from his own measurement τ_2 by

$$\tau_1 = \tau_2/\gamma = \tau_2\sqrt{(1 - V^2/c^2)}. \tag{4.5}$$

Conversely, consider now the same process that takes place in S between the events $(x = 0,\ t = 0)$ and $(x = 0,\ t = \tau_1)$, i.e., the time of the process is τ_1. In S', using (4.1) and (4.2), one finds that the end of the process is measured in S' at $(x' = -\gamma V \tau_1,\ \tau_2 = \gamma \tau_1 > \tau_1)$. One notes that the time measured in the system in which the process occurs is shorter than that measured in another system moving relative to the first.

Consider again the two systems S and S'. Suppose that the same process takes place in both systems. At $t = t' = 0$ the two origins O and O' coincide and this is the first event in the systems: $(x = 0,\ t = 0)$ and $(x' = 0,\ t' = 0)$. An identical process takes place at O and O' and its duration is τ in the both fames. The second event (the end of the process) is in S at $(x = 0,\ t = \tau)$ and in S' at $(x' = 0,\ t' = \tau)$. In S an observer measures τ for the process in his own system (S) and $\gamma \tau$ for the process in S'. Reciprocally, in S' an observer measures time τ for his own process and $\gamma\tau$ for the process in S. In conclusion, these examples show that there is a complete reciprocity of events in the Special Relativity.

A third case is interesting to investigate since it corresponds to the situation of the Clock paradox. The system S' moves away from S at the velocity V and one measures the position of O' in the system S. The first event is defined by $x = x' = 0$ and $t = t' = 0$. The second event in S is as follows: the origin O' travels a distance L during a time equal to L/V, i.e., the event is defined by $(x = L, t = L/V)$. What are the coordinates of this second event measured in S'? From (4.1) and (4.2), one obtains $(x' = 0, t' = L/(\gamma V))$. x' is zero since it is the position of O' in S', and the time measured in S' is shorter than the time measured in S.

Now, one considers the same process but from the point of view of S' in which an observer measures the position of O that goes away with the velocity $-V$. In S', at the end of the process, the position of O is given by the distance $-L'$ and the time is L'/V. In S, using (4.3) and (4.4), one obtains $x = 0$ and $t = L'/(\gamma V)$. What is the relationship between the two distances L and L'? Since there is a complete symmetry between the two points of view of S and S', the times as measured in S and S' must be equal. It results that $L = L'$.

4.2.2. *Simultaneity and length contraction*

Consider two events that occur in S' at two different places, x'_1 and x'_2, but at the same time t': they are what we call simultaneous events. How are these two events seen in the frame S? Using equations (4.3) and (4.4),

$$x_1 = \gamma(x'_1 + Vt') \qquad \text{and} \quad x_2 = \gamma(x'_2 + Vt'), \tag{4.6}$$

$$t_1 = \gamma(t' + Vx'_1/c^2) \quad \text{and} \quad t_2 = \gamma(t' + Vx'_2/c^2). \tag{4.7}$$

From (4.7), one obtains

$$t_2 - t_1 = \gamma(x'_2 - x'_1)(V/c^2). \tag{4.8}$$

The time interval between the two events as measured in S is different from zero. In other words, events that are simultaneous in one frame are not simultaneous in another frame. It is only if the two events are in the same place in S' that they are also simultaneous in S.

Until now, we were interested only in the situations where time is involved. Now we consider the measurement of length from frame to frame. Two points in S' (x'_1 and x'_2) are separated by a distance l_0. This length is the proper length of the distance between the two points. We want to know what is the length between the two points, as measured by the S observer.

This measurement must be done by measuring the positions of the two points at the same time in S. From (4.3) and (4.4), one can write

$$x_1 = vt + x'_1\sqrt{(1 - V^2/c^2)}, \tag{4.9}$$

$$x_2 = vt + x'_2\sqrt{(1 - V^2/c^2)}, \tag{4.10}$$

and finally,

$$x_2 - x_1 = (x'_2 - x'_1)\sqrt{(1 - V^2/c^2)} < l_0. \tag{4.11}$$

The distance between the two points, as measured by the S observer, is smaller than its proper length in S'. This is called the "length contraction".

We saw above a complete reciprocity of the time dilation, i.e., the observer in the frame S notes that the clocks of the frame S' are behind his own clocks, and reciprocally, the observer in S' notes that the clocks of the frame S are behind his own clocks.

Similarly, the observer in S notes that the length of a rod located in the frame S' is shorter than the length of the same rod measured in the frame S'. Reciprocally, the observer in S' notes that the length of a rod located in S is shorter than the length of the same rod measured in the frame S.

One can say that this is something analogous to a parallax. Some authors (Bohm, 1965) make a comparison with the observation of objects at some distance. Two observers, A and B, are located at some distance to one another. A notes that B is smaller than if B were near, and conversely, B notes that A is smaller than if A were near. This analogy gives the possibility of understanding that the reciprocity is not a paradox.

In order to make it clear, we shall call the time of an event in the frame S' measured by one observer of another frame the "measured time" or the "read time" to distinguish it from the proper time of this frame. The proper time is the time given by the clocks in the frame S', since by definition the clock that gives the time is located without velocity in the frame S'.

It is evident that the Relativity Theory is not interested in the age of the observers. However, in the Twin paradox this notion of age is introduced, and we have to recall that the proper time in a frame is also the age of the observers. It is a trivial recall, but as we shall see later, it is not evident for all.

4.3. Dynamics

The fundamental law of dynamics is the second law of Newton for a particle moving along the x-axis under the influence of force F:

$$d(mv)/dt = m(dv/dt) = F, \tag{4.12}$$

where m is the mass of the particle and v its velocity. If the force is constant, the velocity will increase indefinitely for infinite time. But this is in contradiction with the law of Relativity that states that the largest velocity is c, the light velocity. Consequently, this dynamics law must be modified. It is proposed that

$$\frac{d}{dt}\left(\frac{mv}{\sqrt{1-\frac{v^2}{c^2}}}\right) = F \tag{4.13}$$

from which one obtains (F is constant)

$$v = \frac{Ft}{m\sqrt{1+\left(\frac{Ft}{mc}\right)^2}}. \tag{4.14}$$

Taking simple numerical values ($F/(mc) = 1$), one can calculate $v(t)$ and it can be seen in Fig. 4.1 that for $t \to \infty$, the velocity of the

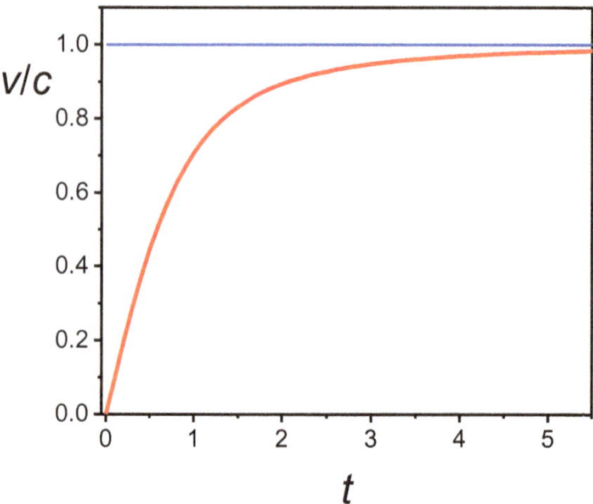

Fig. 4.1. Velocity of a particle under the influence of a constant force. For $t \to \infty$, the velocity tends to the light velocity.

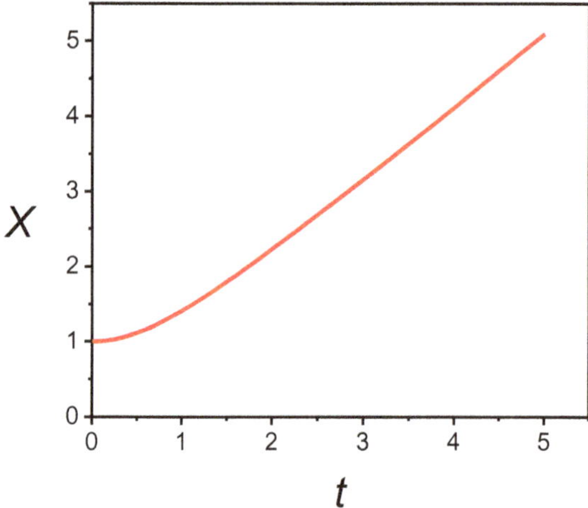

Fig. 4.2. Position of the particle under the influence of a constant force.

particle tends to the light velocity in accordance with the Postulate of Relativity.

Once the velocity is calculated, one can obtain the variations of the absciss a x with t and it is possible to see in Fig. 4.2 that for

$t \to \infty$, $x(t)$ tends to increase linearly since its velocity tends toward the constant light velocity c.

The Langevin Experiment or the Paradox

4.4. The Formulation

What exactly is the Clock paradox or the Twin paradox? It is very often formulated as follows. Consider two twins, Anna and Robert. Anna, who is a very courageous person, decides to make a trip toward a star, since she was able to build an engine, which can travel with a velocity close to the light velocity. She gets very quickly a good velocity V. For Robert, who remains on the earth, the time taken to reach the star is equal to $t_R = L/V$ where L is the earth–star distance. But when Robert wants to know the time taken by Anna to reach the star, he uses Lorentz transformation and finds out that her time is $t_A = L/(V\gamma)$ with $\gamma = (1 - V^2/c^2)^{-0.5}$. When she comes back, the total time for Robert is $2t_R$ and that for Anna is $2t_A$. She is younger than Robert, since $t_A < t_R$.

But the story is not yet finished. Since in Special Relativity, motion is relative, it is possible to view Anna as at rest and Robert as traveling. In this case, Anna notes that Robert goes away and comes back. Now, the travel time of Anna is $2L/V$ and that of Robert is $2L/(V\gamma)$. Consequently, he is younger. How is it possible that Anna is, at the same time, younger and older than Robert? The same thing for Robert.

This simple formulation of the paradox is the age difference that the theory predicts when there is a complete symmetry between the two observers, Anna and Robert. However, one needs a more precise analysis.

We begin with two frames, S and S', and recall the results we obtained above. The two frames move with velocity V relative to one another and they do not stop just as in Special Theory of Relativity. An observer in S, Robert measures the position of the origin of the coordinate axes O' at two events, which form a process. The first event is defined by $x = x' = 0$ and $t = t' = 0$. The second event in S is as follows: the origin O' travels a distance L during a time equal to L/V i.e., the event is defined by $(x = L, t = L/V)$. What are the coordinates of this second event measured in S'? From the Lorentz transformation, one obtains $(x' = 0, t' = L/(\gamma V))$. x' is zero since it is the position of O' in S', and the time measured in S' is shorter than the time measured in S. Note that S' does not stop after traversing the distance L.

What is the meaning of the coordinates $x' = 0$ and $t' = L/(\gamma V)$. They are the coordinates in S' of an event that occurs in the system S. The process of motion of O' from $x = 0$ to $x = L$ does not concern S'. During the process, S' is at rest with itself. This means that the time $t' = L/(\gamma V)$ is not the proper time of S' and is not the age of Anna.

Now one considers the same process but from the point of view of S', in which Anna measures the position of the origin of the coordinates O that goes away with velocity $-V$. In S', at the end of the process, the position of O is given by the distance $-L$ and the time is L/V. In S, using again the Lorentz transformation, one obtains $x = 0$, and $t = L/(\gamma V)$. From this, it can be seen that during the process seen by S' and S, the proper times of Anna and Robert are the same. They have the same age. The two frames do not stop and there is no paradox. However, in the Langevin experiment, one of the frames stops and we shall consider what happens during the stopping.

A curious result is that when Anna considers the travel of Robert, she notes L/V, but when she considers the motion of her own coordinates origin, she notes $L/(\gamma V)$. This is because these are two different processes in two different frames. The same thing for Robert.

Now what happens if Anna in S' decides to activate her engine just before O' reaches the distance L so that the relative velocity between the frames becomes zero? One can say that Anna enters the frame of Robert. In this new state, they can compare directly their clocks. Because of the reciprocity of the motion, one may conclude that it is Robert who enters the frame S'. Consequently, one expects that their clocks will indicate the same time. However, the application of the theory shows that Anna is younger. This is the paradox, and the goal of the theory is to show that the two twins do not have the same age.

4.5. When a Frame Stops Its Motion

For this, one has to examine a problem for which one does not find the solution in any textbook. What happens to the clock of a frame which decreases its finite velocity V to a velocity null relative to another frame?

Until now, we considered only the inertial frames with constant velocity. The observers of the different frames never meet one another and they remain in their own frames. This was also the case in the first part of the Langevin article. However, in the second part, he introduces acceleration and considers the meeting of the observers on the earth on the return from the trip. At the encounter, they can compare their proper times, which are, as recalled above, their ages. We shall try to retrieve the result of Langevin by the use of an invariant.

First, we define the squared space–time distance between two events (x_1, t_1) and (x_2, t_2) as

$$D^2 = c^2(t_2 - t_1)^2 - (x_2 - x_1)^2. \tag{4.15}$$

It is possible to verify that between two events, the space–time distance is an invariant, i.e., it takes the same value in all inertial reference frames. If the squared space–time distance is negative, one has what Langevin calls a *pair in space* and if the squared space–time distance is positive, these events form a *pair in time.*

We consider again the two frames S and S', but now the frame S' is not an inertial one; this means that its velocity relative to S is not constant. As mentioned above, the two frames move (one relative to the other) along the x-axis. A process occurring in S' can be seen in S as a series of successive events located very close to one another in their space–time distances. To each of these intermediate events one associates an instantaneous inertial frame. At each instant t (in the inertial frame), the process is characterized by an event located at x. Between two very close events (t, x) and $(t + dt, x + dx)$, the quantity D defined as

$$D^2 = c^2(dt)^2 - (dx)^2 = c^2[(dt)^2 - (dx/c)^2]$$
$$= c^2(dt)^2[1 - (dx/dt)^2/c^2]$$

or

$$D = cdt\sqrt{\left(1 - \frac{V(t)^2}{c^2}\right)} \quad (V(t) = dx/dt)$$

is an invariant, i.e., it takes the same value in all inertial references. We define the instantaneous proper time $d\tau$ of the accelerated frame as

$$d\tau = dt\sqrt{1 - \frac{V(t)^2}{c^2}}. \tag{4.16}$$

This particular time defines the proper time of an inertial reference frame that is moving with velocity $V(t) = (dx/dt)$ and with the origin located at the point x. We saw that the process is associated with a succession of inertial frames and in fact, they can be seen as the successive positions of inertial frames which describe the motion of the non-inertial frame S'. If one considers the whole process between two times t_1 and t_2 and between two positions x_1 and x_2 (in the frame S), the proper time of S' is given by

$$D\tau = \int_{t_1}^{t_2} dt\sqrt{1 - \frac{V(t)^2}{c^2}}. \tag{4.17}$$

This proper time is independent of the frame in which the measurement of t_1 and t_2 was made. In other words, in another frame different from the first one, the calculated proper time is still the same. Expression (4.17) opens the possibility of calculating the proper time of some process in any frame one chooses.

In the case where (dx/dt) is constant (where S' is now an inertial frame), the process takes place in the same place in an inertial reference frame that moves with velocity $V = (dx/dt)$. In such a case, for the proper time of S', one has merely

$$D\tau = (t_2 - t_1)\sqrt{1 - \frac{V^2}{c^2}} = (t_2 - t_1)/\gamma \qquad (4.18)$$

with $\gamma = \frac{1}{\sqrt{(1 - V^2/c^2)}}$ and $(t_2 - t_1)$ is the time elapsed for the process in the original inertial frame. One finds once again the above result that the time of the process in the frame in which it occurs (here S') is shorter than that measured in another frame moving with velocity V.

The expression (4.17) is very important, as we shall see below and later. It seems that the introduction of the frame S', which is not an inertial frame, is not compatible with the theory of Special Relativity. From the Einstein paper of 1905 and textbooks, there is apparently no place for accelerated frames; it is the domain of General Relativity. However, we introduced it here, just as some textbooks (Bohm, 1965; Møller, 1952) did, because it is only some extension of the theory and it does not need the new concepts developed in General Relativity.

Now, one supposes that in the frame S', which until now was inertial, there is a motor engine that is able to change its velocity relative to S. Anna in the frame S' decides to activate her motor engine such that her velocity relative to S is modified and becomes zero. In other words, at the end of this process, the frames S and S' become the same reference frame. The velocity of S' relative to S is equal to V between $t = 0$ and t_0, and for $t > t_0$ it is a decreasing function $f(t)$ so that for the time t_1, $f(t_1) = 0$. To fix the things, suppose that the two clocks were synchronized such that $t = 0$ corresponds to $t' = 0$. The people in S' activate their motor engine at the time t_0 measured in S and their relative velocity is null at the time t_1 measured also in S. The question arises: what are the indications of the two clocks when they meet in S?

The rate of a clock is difficult to define or to measure. What can we say if the rate of the clock A is larger or smaller than the rate of the clock B? Consider two clocks at rest in two inertial frames, S and S'. In both frames occurs the same event dependent only on time such as a chemical reaction. In both frames, the reaction will take time τ and we shall say that the two clocks have the same rate. If S' moves with velocity V, for the observers in S, the clocks of S and S' do not have the same rate, since for the observers in S, the reaction in S' will take time $\gamma\tau$. Before we answer the above question, it is clear that the two clocks (that of S and S') will have the same rate when reunited.

In S, we note the time of any event indicated by the clock as merely t_1, and in S', the clock will indicate the proper time of the frame S'. It is given by (4.17), where the velocity is measured in S. One has, for the time t_1' indicated by the clock of S',

$$t_1' = \int_0^{t_1} dt\sqrt{1 - \frac{V(t)^2}{c^2}} = \int_0^{t_0} dt\sqrt{1 - \frac{V^2}{c^2}} + \int_{t_0}^{t_1} dt\sqrt{1 - \frac{f(t)^2}{c^2}}.$$

$$(4.19)$$

After the time t_1, the S' clock has the same rate as the S clock but the two clocks are not synchronized: the difference in the indications of the two clocks is precisely $\delta = t_1' - t_1$. We note that if $\delta = 0$, this means that the two are synchronized (the same time indication and the same rate).

In order to calculate the time t_1' from (4.19), we shall use an approximate method. For this, we shall use the following theorem.

Suppose, for a continuous function $f(x)$, there are two values, m and $M(m \leq M)$, such that

$$m \leq f(x) \leq M.$$

Thus, there are two inequalities concerning the integral $\int_a^b f(x)dx$:

$$m(b - a) \leq \int_a^b f(x)dx \leq M(b - a).$$

We shall take as an approximate value of the integral, the means of the two sides of the inequalities and we write

$$\int_a^b f(x)dx = \left(\frac{M + m}{2}\right)(b - a).$$

We apply the above expression to the second integral of (4.19) and we obtain

$$\int_{t_0}^{t_1} dt \sqrt{1 - \frac{f(t)^2}{c^2}} = \frac{1+k}{2}(t_1 - t_0)$$

with $k = \sqrt{(1 - V^2/c^2)}$. From (4.19), we obtain

$$t_1' = \frac{t_0}{2}(1 - k) + \frac{t_1}{2}(1 + k).$$

It is not difficult to show that t_1' is shorter than t_1 and the difference δ is not null. For this, one writes $t_1' < t_1$.

$$t_1' = \frac{t_0}{2}(1 - k) + \frac{t_1}{2}(1 + k) < t_1,$$

$$\frac{t_0}{2}(1 - k) < \frac{t_1}{2}(1 - k).$$

Since $t_0 < t_1$, the inequality $t_1' < t_1$ is verified. The clock of S', once it enters the frame S, is behind the clock of S.

One quotes the following words of Campbell (1940):

> The special theory of relativity deals with uniform relative motion and a body under such motion cannot make a 'return trip' from another body. But I have made an extension [Campbell 1933] of the Lorentz transformation to accelerated motion, and this extension makes it possible to consider a return trip. The validity of this extension is confirmed by the consistency of the results from the point of view of the observers in the two systems, and also by their agreement with the treatment of time distortion by the general theory of relativity. The conclusion is that while the rod on returning has the same length as the one which remains stationary, the total recorded time of the moving system is less than the total recorded time of the stationary system; that is, the time distortion is real.

It is interesting to report a discussion between a student and Einstein. In 1911, Einstein gave a lecture in Zurich on relativity. I quote Pesic (2003):

> During the subsequent discussion, a young law student named Fritz Muller raised a question about the clock example: 'According to the

reasoning given in the lecture, at the moment of its meeting with the other clock at point A, the second clock will not be running in synchrony again. How can this be possible, since, on the other hand, Professor Einstein says that a rod of a specific length L in a system at rest, which he holds in his hand, will become shorter by a definite amount when it is set in motion? But as soon as the rod is brought to halt by a sudden jerk, its length is once again $= L$, i.e., the rod is no longer deformed'. Muller goes on to apply this to the clock: 'Must it not then, exactly like the rod, run synchronously again from the moment it is brought to rest at point A?'

Einstein's wonderfully clear reply shows again the clarity of his understanding that the real issue is simultaneity: 'It is not the clock's indication of time that is to be likened to the rod, but its rate. After having completed its motion and returned, the rod has the same length. In the same way, the clock has again the same rate. We can designate the rod as the carrier of the space differential, and the clock as the carrier of the time differential. It is impossible to assume that, after having traveled along a polygonal path and returned to point A, the clock will again be running synchronously with the clock that has been at rest at point A.

This is the result given by Langevin. After the application of acceleration to S', upon the encounter of the two frames, the proper times are not equal; in the frame S' (which was accelerated but is now in the frame S), the clocks are behind the clocks of S. One can understand why Langevin needed acceleration. In the absence of acceleration, the two observers cannot meet and it is only upon acceleration that the velocity of the traveler becomes zero relative to the stay on the earth. Now, a meeting is possible and the comparison of their ages becomes meaningful.

In other words, if the two frames are moving one relative to the other, the measured time from frame to frame is a kind of parallax. However, if the two frames have together formed one unique frame, the time difference appears as a real effect given by the differing readings of the S clock and S' clock, despite being located in the same frame and having the same rate.

4.6. A Graphical Solution

We sum up the Langevin problem by a graphical representation shown in Fig. 4.3, considering only the outward travel. The return is identical to the outward travel. The x-axis gives the time t of the stay at home and the y-axis gives the times t and t': the time t is the proper time of the stay at home and t' is the time of the traveler. The red line is the proper time of the stay at home and is such that $x = y$. The blue line gives the time of the traveler.

From $t = 0$ to t_1, the two observers are at rest in the frame of the stay at home, the red and blue lines coincide, and the clocks are synchronous. From t_1 to t_2, the traveler begins his motion and he is accelerated. From t_2 to t_3, the velocity of the traveler is constant and he measures his time as t/γ. From t_3 to t_4, he is decelerated until his velocity relative to the stay at home is null. For $t > t_4$, the two observers are now in the same frame. Their clocks have the

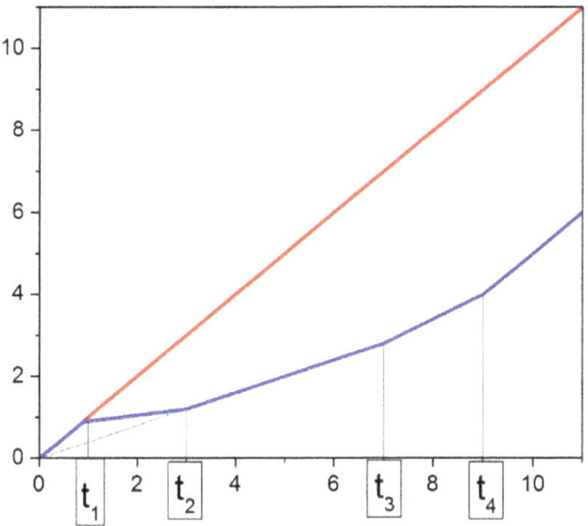

Fig. 4.3. The x-axis is the time of the stay at home and the red line (with slope equal to 1) is also the time of the stay at home. The blue line is the time of the traveler. Between $t = 0$ and t_1, the two frames coincide. Between t_1 and t_2, the traveler is accelerated. Between t_2 and t_3, he has a constant velocity and the blue line is a straight line with slope t/γ. Between t_3 and t_4, he is decelerated and for $t > t_4$ the two observers are in the same frame with clocks having the same rate.

same rate (the red and blue lines are parallel) but indicate different times, with that of the stay at home being longer than that of the traveler.

One can make two remarks:

If $t_3 = t_4$ (immediate stopping), one finds that the ratio of the times indicated by the clocks of the two frames is equal to γ.

If t_4 goes to infinity (very slow stopping), the times of the two clocks tend to be equal.

This surprising result is very difficult to accept, because one does not see the relationship between the cause (here, the acceleration) and the consequence (the age difference). In several physical phenomena, there is a clear connection between the cause and the consequence. Take as an example the Joule effect in metals. The process takes place as follows. On the application of an electric field to a piece of metal, the free electrons begin to move all in the same direction (in the mean), and by their collisions with the metal atoms, they transmit their electric energy to the lattice. Consequently, the metal becomes hotter. The relationship between the applied voltage (which creates the electric field) and the heating of the metal is clear. However, this is not the case with the age difference. As we shall see, the General Theory of Relativity gives the answer.

General Relativity

The General Theory of Relativity is much more complex than the Special Theory of Relativity, and in particular, the mathematics are at high level, being well above those necessary to understand Special Relativity. We recall here only same notions, which are needed in order to understand the Langevin problem. They are three steps towards a qualitative approach to General Relativity.

4.7. The Two Aspects of the Concept of Mass

The concept of mass appears in two different contexts. In the first, there is attraction because of the phenomenon of gravity. This law is well known, and if one considers a particle of mass m_1 and another of mass M, the force between them is given by

$$F = G\frac{m_1 M}{R^2}, \tag{4.20}$$

where R is the distance between the two masses and m_1 is the gravitational mass. The second situation is when a force F is applied to the same particle. The second law of Newton gives

$$F = m_2(dV/dt). \tag{4.21}$$

V is the velocity of the particle. We denote the mass by m_2, because it is an entirely different situation: it is the inertial mass. In (4.20), the force appears because of the presence of the mass M. In (4.21), the mass is a parameter that describes the opposition of the particle against the motion produced by the applied force.

Although we used to say that m_1 is equal to m_2, i.e., there is only one mass, the same mass, this equality is not evident and apparently there is no reason for it. A simple method to check this equality is using the free fall of objects on the earth's surface. In such a case, equation (4.20) reduces to $F = m_1 g$, and the free fall equation is given by

$$F = m_1 g = m_2(dV/dt). \tag{4.22}$$

If, as shown by experiments, the fall is independent of the mass of the particle, this means that $m_1 = m_2$.

4.8. The Gravitation–Acceleration Equivalence

The equality of the inertial mass to the gravitational mass induced Einstein (1911) to propose a very important notion: the principle of equivalence. It can be formulated by stating that the consequences of a gravitational field on a mass are equivalent to the consequences of an acceleration applied to that mass.

To understand intuitively this principle, one considers two identical systems made of two identical boxes with two observers in each box. At the roof of each box, each observer hangs a spring with a mass. Now, one places one box in a gravitational field such that the mass tends to go down and the observer notices that the string is elongated downward. Of course, this system is clamped such that it does not move. One gives to the second box an acceleration relative to a frame at rest such that this box moves upward. Here also the spring is elongated downward. If the two observers cannot see what is outside, they cannot decide whether the elongation of the spring is due to a gravitational field or an acceleration. This is the principle of equivalence.

This principle will be used to establish the properties of the gravitational field from the properties of acceleration and vice versa.

4.9. The Time in a Non-inertial Frame

From the principle of equivalence, a non-inertial frame is an accelerated frame or a frame in a gravitational field. An important notion is that under a gravitational field, the rate of time is different from that in a frame without a gravitational field.

This effect is not a parallax effect as the time dilation in Special Relativity but instead an intrinsic effect. It is the key necessary for the explanation of the Langevin experiment.

To understand this concept of time dilation in General Relativity, we shall follow Einstein who proposed it in his paper of 1911. One considers two reference frames xyz and in each of them two points on the z-axis, S_1 and S_2, such that the z coordinate of S_2 is higher than that of the point S_1. In one of the frames, there is a gravitational field such that the potential of S_1 is taken as the reference point (its potential is null) and the potential of S_2 is higher than or equal to $h\gamma$ when h is the distance between the two points S_2 and S_1. The second frame is accelerated upward with the acceleration γ.

In the accelerated frame, at the point S_2, there is a light emitter that emits light with frequency ν_2. At the point S_1 of this frame, there is a receptor that receives light with frequency ν_1. The frequencies ν_2 and ν_1 are not equal. When some periods of light are emitted, because of motion of the frame upward the receiver receives a larger number of periods within the same time. This is the Doppler effect and in the classical approximation, one has

$$\nu_1 = \nu_2(1 + V/c), \tag{4.23}$$

where V is the velocity of the frame and c is the light velocity. The velocity V is given by the equation $v = \gamma t$ when t is the time for the light to pass from the emitter to the receiver. One has $t = h/c$ and $V = \gamma h/c$. Expression (4.23) becomes

$$\nu_1 = \nu_2(1 + \gamma h/c^2). \tag{4.24}$$

Now, one translates the emitter and the receiver to the second frame in the gravitational field and by the principle of equivalence, one can say that again in this frame the frequencies ν_1 and ν_2 are related by the expression (4.24), which can be written as

$$\nu_1 = \nu_2(1 + \Phi/c^2), \tag{4.25}$$

where $\Phi = h\gamma$ is the gravitational potential.

Now Einstein asks a very important question:

On a superficial consideration, equations (4.24) or (4.25) respectively seems to assert an absurdity. If there is a constant transmission of light from S_2 to S_1, how can any other number of periods per second arrive in S_1 than is emitted in S_2?

For Einstein, the answer is very simple:

Nothing compels us to assume that the clocks U [in the vicinity of the points S_2 and S_1] in different gravitation potentials must be regarded as going at the same rate.

In order that both clocks, the clock near S_2 and that near S_1, count the same number of periods within a given time, it suffices that the clock in S_2 has a different rate by an amount equal to $(1 + \Phi/c^2)$.

To be intuitively more precise, suppose that the emitted light emits 10 periods per second and that the light is received at 12 periods per second. In order that the clock of S_2 counts 12 periods per second, it is necessary that it goes slower by a factor of $10/12$.

The general conclusion of this analysis is as follows: in a gravitational field the clocks go in a different manner than the same clocks in a position without gravitation. In Chapter 5, we show how Einstein uses all these concepts of General Relativity to explain the Langevin problem.

In fact, equation (4.25) is only an approximation, since the formula (4.23) is the classical formula for the Doppler effect. In Special Relativity, the formula for the Doppler effect is different and equation (4.25) is replaced by

$$\nu_1 = \nu_2(1 + 2\Phi/c^2)^{0.5}. \tag{4.26}$$

The final formula for the equation that relates the times of a process that takes the time t_1 without gravitation and the time t_2 with gravitation is

$$t_2 = t_1(1 + \Phi/c^2). \tag{4.27}$$

It is possible to verify equation (4.27) with a simple example of two frames: one frame (S_1) considered to be at rest and a second frame (S_2) moving (relative to S_1) with a constant acceleration k. One considers a process that takes the time τ_1 in S_1 and now wants to know what is the time τ_2 at the end of this process measured in S_2. For this, one uses equation (4.17), which may be written as

$$\tau_2 = \int_0^{\tau_1} d\tau_1 \sqrt{1 - \frac{V(\tau_1)^2}{c^2}}. \tag{4.28}$$

One supposes that for $\tau_1 = \tau_2 = 0$, the origins of S_1 and S_2 coincide. In such a case, the velocity of S_2 relative to S_1 is $V = k\tau_2$ and equation (4.28) becomes

$$\tau_2 = \int_0^{\tau_1} d\tau_1 \sqrt{1 - \frac{(k\tau_1)^2}{c^2}}. \tag{4.29}$$

One makes normalization of the variables in writing

$$t_1 = k\tau_1/c = \tau_1/\tau_0,$$
$$t_2 = k\tau_2/c = \tau_2/\tau_0$$

and equation (4.29) now becomes

$$t_2 = \int_0^{t_1} dt_1 \sqrt{1 - t_1^2}. \tag{4.30}$$

The time $\tau_0 = c/k$ is the time for which the velocity of S_2 is equal to c; this means that it is the largest possible time. Consequently $t_1 \leq 1$ and $t_2 \leq 1$.

The integration of (4.30) yields

$$t_2 = 1/2 \left[t_1 \sqrt{1 - t_1^2} + Atg \left(\frac{t_1}{\sqrt{1 - t_1^2}} \right) \right]. \qquad (4.31)$$

From (4.31), one can draw the curve of the ratio (t_2/t_1) as a function of t_1. (see Fig. 4.3). One sees that for $t_1 = 0$, the ratio $p = t_2/t_1$ is equal to 1, i.e., the proper times are equal as expected. However, when t_1 increases, the ratio p decreases, indicating that the rate of time for S_2 is slower than that for S_1.

Now, one tries to fit this curve with the expression (4.27) and so one writes

$$t_2/t_1 = (1 + \Phi/c^2). \qquad (4.32)$$

To determine the gravitational potential, we need to consider the situation from the point of view of S_2. From this frame, the frame S_1 moves toward the negative x and the potential Φ is equal to $-kx$, where x is the distance between the frames. In the approximation of the Newtonian mechanics (which will be good for small velocities or small times), one has $x = 1/2kt_1^2$. Consequently, one has to fit the ratio p with the function

$$(1 - At_1^2), \qquad (4.33)$$

where A is a constant. We already know that we cannot expect a good fit for all the values of t_1 because of two approximations. First, because the expression (4.32) is based on the classical Doppler effect, and secondly, because the equation $V = kt$ is good only for small times. A more exact calculation should use the expression (4.14) for the relationship between V and t, but it is more complicated. Effectively, one sees in Fig. 4.4 that for $t_1 < 0.3$ the fit is very good but not for larger values.

We found that the preceding result, namely that an observer who moves with a variable velocity (under the application of an acceleration) will be younger than an observer at a constant velocity, was already applied to the Clock paradox. We bring two articles on this subject: the author of the first is Lass (1963) and the authors of the second are Gambon *et al.* (2018). Of course, it is likely that

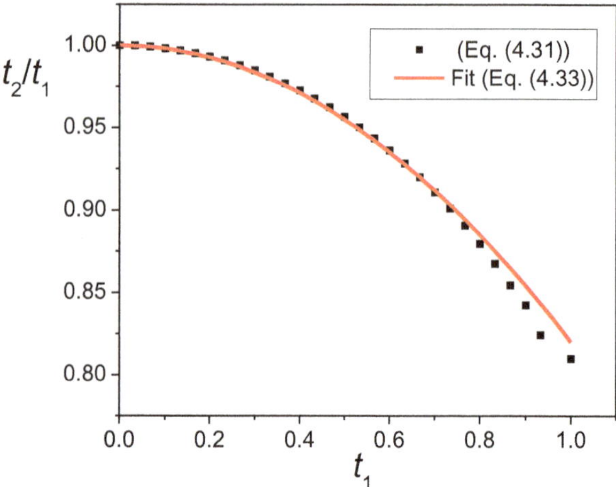

Fig. 4.4. Ratio of the proper times of the two frames: Frame 1 is inertial and frame 2 is accelerated.

we missed other publications. It is worth quoting the last lines of the abstract of the latter authors:

> In the resolution of the Twin paradox, the role of the acceleration has been denigrated to the extent that it has been treated as a red-herring. This is a mistake and shows a clear misunderstanding of the Twin paradox.

The theory of General Relativity develops other concepts but we do not intend to discuss them in the frame of this book. In particular, the theory describes space–time as a non-Euclidian curved space. But there is no need for this curved space in the analysis of the Langevin experiment.

4.10. The Langevin Problem

At this point, one can understand this sentence of Langevin:

> We can still say that it is sufficient to be agitated or to undergo accelerations, to age more slowly, and we'll see in a moment how much you can expect to get in this way.

The basic explanation of the age difference between the two observers is as follows (following Einstein): the traveler lived (during the time of application of the acceleration) more slowly relative to a clock at rest. The age difference effect is related to the properties of bodies in acceleration: it is a real effect and not a parallax. This gives the qualitative explanation of the Clock paradox. In the first section, we saw that the stopping induces the age difference. Here it is directly related to the acceleration.

Chapter 5

The First Solutions

In this chapter, we report solutions that are probably among the first published. Surprisingly, these solutions present the characteristics of several solutions presented later on during the 20th century.

5.1. The First Einstein Argument

In his first paper on Relativity in 1905, Einstein imagines that a clock moves with a constant value of velocity along a broken closed line. The clock begins its motion at the point A of this closed line and comes back to the same point A, where another clock remains at rest. Einstein concludes that the two clocks (which were synchronized at the beginning of the motion) are no longer synchronized and the clock that moved is behind the clock at rest.

This statement of Einstein is a real surprise and one can ask at least two questions. The first concerns the path of the moving clock. Since it is a polygonal curve, at each breaks, one needs to apply a force that results in acceleration. Is the theory able to deal with acceleration? In the 1905 article, there is no mention of acceleration. One can ask a second question as follows: does the moving clock stop at its return to the point A or continue its motion? There is an ambiguity that Einstein does not help us to resolve. Following the content of the paper, the answer to this second question is: the clock continues its motion since in no place in the paper the stopping of a clock is evocated. But in fact it seems that Einstein intends the first answer: the moving clock stops at the return point. We quote Pesic (2005) again:

> Consider his statements in a speech he gave in 1911 to the Naturforschende Gesellschaft in Zurich, "which included his presentation of what he called 'the thing at its funniest' (*am drolligsten*), the two clock example much as in the 1905 paper. He extends the example to twins: 'Were we, for example, to place a living organism in a box and make it perform the same to-and-fro motion as the clock discussed above, it would be possible to have this organism return to its original starting point after an arbitrarily long flight having undergone an arbitrarily small change, while identically constituted organisms that remained at rest at the point of origin have long since given way to new generations'.

From this quotation, it is clear that, for Einstein, the clock that moves on the closed line stops at the end of its path. This is why frequently it is claimed that the precursor of the Twin paradox is Einstein himself and not Langevin.

But in fact, there is a very big difference. For Einstein, the time difference of the clocks is self-evident, but for Langevin, it is the consequence of the application of an acceleration. Einstein does not provide any explanation for his conclusion and apparently, there is confusion between the parallax effect and the result of stopping. This ambiguity is rarely mentioned and one can quote an article by Essen (1971) who notes it. It seems that this confusion of Einstein was also noted by Landau and Lifshitz, who wrote in their book *The Classical Theory of Fields* about the Clock paradox thus:

> If we have two clocks, one which describes a closed path returning to the starting point (the position of the clock which remains at rest), then clearly the moving clock *appears* to lag relative to the one at rest.

The essential word in this quotation is "appears". This means that the moving clock does not stop at the returning point, in agreement with the article of Einstein in which in no place the stopping is mentioned.

But in every case, Einstein was right because effectively at the stopping, the moving clock indicates a shorter time than that of the non-moving clock. But it is an effect due to the principle of equivalence. If one takes the limit of the polygon as a circle, there is a centrifugal acceleration on the particle and consequently the times of the two clocks are not equal. This effect is well known and is analyzed in several textbooks (Arzeliès, 1966; Tonnelat, 1959; Moller, 1952) as the rotating disk.

5.2. The Lorentz Solution

Lorentz proposed in 1914 a solution that was published in a French journal *La Revue Générale des Sciences*. It is probably the first solution published in a scientific journal. It seems that this article was not very well known, which maybe because it was written in French. The article may be considered as an introduction to the Special Theory of Relativity: the title is *Considératins élémentaires sur le principe de relativité* .

As was so frequently assumed, the return of the traveler is very quick, i.e., the acceleration time is very short. To calculate the travel times of the two observers, at the time t_1 (measured in the stay at home frame), the traveler B sends an electromagnetic signal to the stay at home A, which is reflected to the traveler at the time t_2 (always measured in the frame of A).

Lorentz distinguishes three periods in the process. In the first part, the traveler goes away from A at t_1 and at t_2 (t_1 and t_2 are smaller than T, which is half of the total traveler time in this frame). In the second part, at t_1 the traveler goes away, but at t_2, he is coming back and finally in the third part, at the times t_1 and t_2 he is coming back (see Fig. 5.1).

The travel time of the traveler in his own frame is given by

$$t' = \frac{(t_1 + t_2)}{2\gamma} \tag{5.1}$$

and that of the stay at home by τ (see Fig. 5.1)

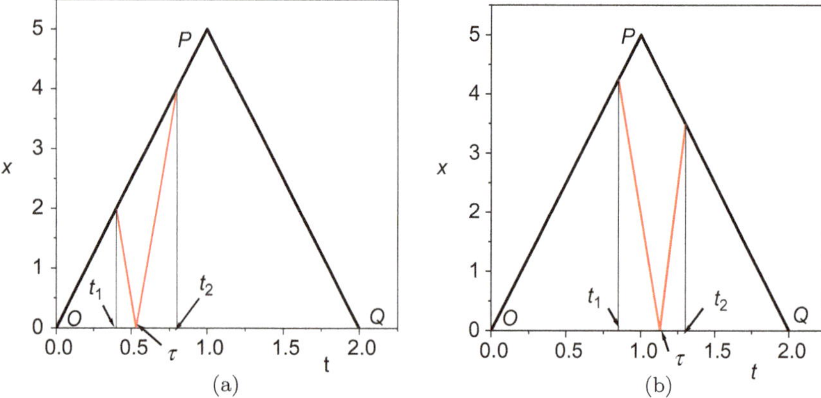

Fig. 5.1. In the plane (x, t), the path of the travel is given by OPQ. At the time t_1, the traveler sends a light signal, which is reflected by the stay at home at the time τ and reaches the traveler at the time t_2. In (a), one has t_1 and t_2 smaller than T and (b), one has $t_1 < T$ and $t_2 > T$.

The complete calculation shows that in the first and third parts, B sees the clock A having a rate smaller than the rate of the clock B, but in the second part, the rate of the clock A increases and compensates the rate in the first and third parts. The exact relations between t' and τ are as follows.

In the first part, which begins at $\tau = 0$ and finishes with $t' = T\sqrt{\frac{c-V}{c+V}}$ and $\tau = \frac{c-V}{c}$, one has

$$\tau = \frac{t'}{\gamma}. \tag{5.2}$$

In the second part, which finishes with $t' = T(1 + \frac{2V}{c})\sqrt{\frac{c-V}{c+V}}$ and $\tau = \frac{c+V}{c}$, one has

$$\tau = t'\sqrt{\frac{c-V}{c+V}} - \frac{VT}{c}. \tag{5.3}$$

Finally, in the third part, one has

$$\tau = \frac{t'}{\gamma} + \frac{2V^2}{c^2}T, \tag{5.4}$$

with the final values $t' = \frac{2T}{\gamma}$ and $\tau = 2T$.

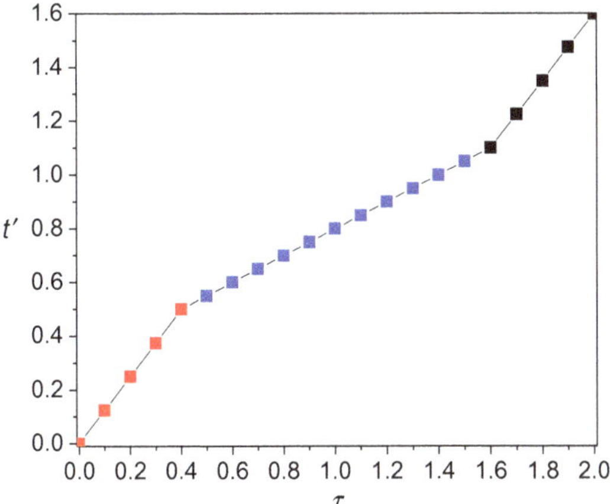

Fig. 5.2. Variation of the traveler time (t') as a function of the time of the stay on earth (τ) in the Lorentz model.

The variation of t' with τ are given in Fig. 5.2. We chose $T = 1$ and $\gamma = 1.25$ ($\beta = 0.6$).

It is interesting to see that Lorentz had some intuition about the possible role of the acceleration. In a note, Lorentz writes (thus translated):

In the first and third periods, the stopwatch of C [that of the stay at home A] goes more slowly for B [the traveler] than its own apparatus ... However, this effect is more than compensated by the acceleration of C relative to C' [the clock of the traveler] which is observed in the second period.

The model of Lorentz is very special because in numerous solutions based on the Special Theory of Relativity, the turnover implies a discontinuity in the time of the stay at home as seen by the traveler. But here the second part of the travel took a long time without discontinuity.

5.3. The Langevin Solution

It has been reported that during Einstein's visit to Paris in 1922, at the *Collège de France*, there was a meeting of physicists which included Langevin (Morand, 1922). Morand, in an article published in the French journal *La Nature* (not to be confused with the English journal *Nature*) tries to expose the solution of Langevin.

He does not consider a traveler launched by a projectile as in his article but a case more realistic of a train with a conductor (the traveler), which makes a trip in front of the station master (the stay at home) who stands on the platform. The first part of the travel is in the $x > 0$ direction.

Langevin begins by recalling the Lorentz equations applied to the problem. During the first part of the travel, we have, for the time t of the station master versus the time t' of the conductor,

$$t = \frac{1}{\sqrt{1 - \frac{V^2}{c^2}}}(t' + \frac{Vx'}{c^2}). \tag{5.5}$$

For $x' = 0$ (position of the conductor), one has $t = \frac{t'}{\sqrt{1-\frac{V^2}{c^2}}}$. Now, for the conductor, one has

$$t' = \frac{1}{\sqrt{1 - \frac{V^2}{c^2}}}\left(t - \frac{Vx}{c^2}\right). \tag{5.6}$$

For $x = 0$ (position of the station master), $t' = \frac{t}{\sqrt{1-\frac{V^2}{c^2}}}$.

The train stops suddenly after it has made the path $x = V\theta$, after a time θ measured in the frame of the station master. One has, for the time of the conductor,

$$t_1' = \frac{1}{\sqrt{1 - \frac{V^2}{c^2}}}\left(\theta - \frac{V^2}{c^2}\theta\right) = \theta\sqrt{1 - \frac{V^2}{c^2}}. \tag{5.7}$$

After the train changes its direction and goes toward the direction $x<0$, the equation giving the time of the conductor would be

$$t_1' = \frac{1}{\sqrt{1 - \frac{V^2}{c^2}}}\left(\theta + \frac{V^2}{c^2}\theta\right), \tag{5.8}$$

but (5.8) and (5.7) must be equal (the conductor keeps the same time when changing his direction), but this is not the case. Then Langevin proposes to add a constant to equation (5.8)

$$t'_2 = \frac{1}{\sqrt{1 - \frac{V^2}{c^2}}} \left(\theta + \frac{V^2}{c^2}\theta \right) + K, \qquad (5.9)$$

such that (5.9) will be equal to (5.7). A simple calculation gives $K = -\frac{1}{\sqrt{1-\frac{V^2}{c^2}}}(2\frac{V^2}{c^2}\theta)$ or $K = -\frac{1}{\sqrt{1-\frac{V^2}{c^2}}}(2\frac{VL}{c^2})$ since $V\theta = L$. Now, one can write

$$t'_2 = \frac{1}{\sqrt{1 - \frac{V^2}{c^2}}} \left[\left(t - 2\frac{VL}{c^2} \right) + \frac{Vx}{c^2} \right], \qquad (5.10)$$

which for $t = \theta$ and $x = L$ gives (5.7).

In other words, for the conductor, the time of the station master has been changed by the quantity $2VL/c^2$. When the train comes back to its initial point, the time of the station master is 2θ. Inserting $t = 2\theta$ and $x = 0$ in (5.10), one finds the time of the conductor

$$t'_3 = 2\theta\sqrt{1 - \frac{V^2}{c^2}}, \qquad (5.11)$$

as expected. It is important to note that Langevin uses the Special Theory of Relativity and does not use the principle of equivalence. Similar to Lorentz (but apparently in a different manner), the age difference appears at the turnover. Langevin does not mention directly the acceleration but the "gravitational field". It is clear that he hints at the principle of equivalence, which puts in parallel the acceleration and the gravitational field.

5.4. The Einstein Solution

In a paper published in 1918, Einstein imagines having a dialogue with Criticus, a physicist who does not accept the conclusions of the Theory of Relativity. This physicist asks questions and Einstein answers them. It is titled "Dialogue about Objections Against the Theory of Relativity".

Several authors (Pesic, 2005; Sachs, 1971; Dingle, 1956) have noted that the explanation proposed in this 1918 paper is different from the simple affirmation of the age difference in the 1905 paper. Should we conclude that Einstein did correct himself? Since Einstein's article in 1918 is not very well known, one finds frequently that the first mention of the paradox is the article in 1905.

As expected, the first question concerns the Clock paradox. Einstein explains the problem in considering two systems K and K'. The system K is that of the observer on the earth and K' is that of the traveler. There is one clock U1 in K and another clock U2 in K'

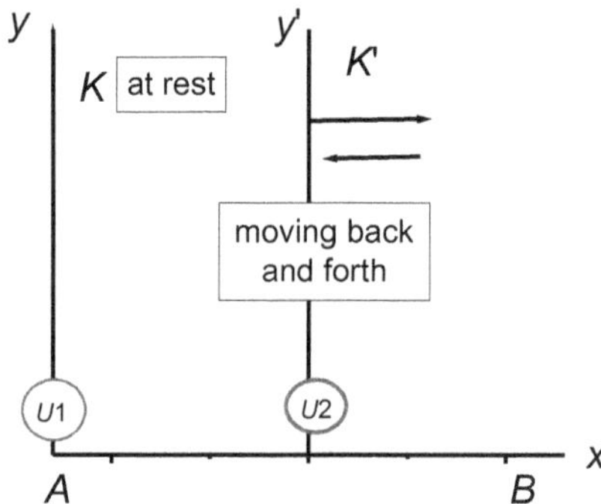

Fig. 5.3. The system K is at rest and the system K' moves from the point A to point B and comes back to A.

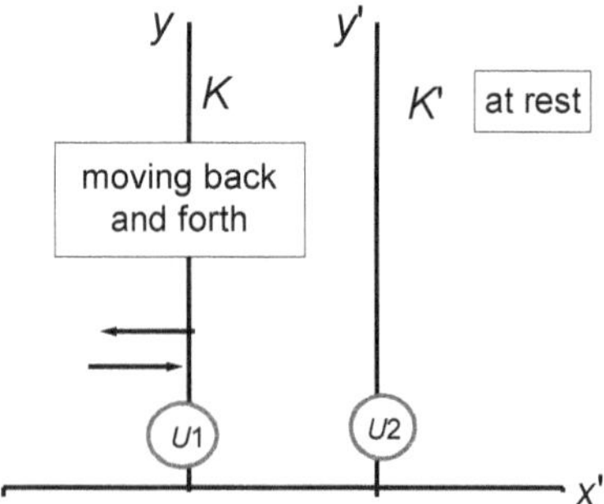

Fig. 5.4. The system K' is at rest and the system K, as seen by the traveler, moves back and forth.

(Figs. 5.3 and 5.4). Now, one quotes Einstein thus:

K is the inertial reference frame.	K' is the accelerated reference frame.
1. The clock U2 is accelerated by an external force along the positive x-axis, until it has reached velocity v. U1 remains at rest.	1. A gravitational field appears, that is directed towards the negative x-axis. Clock U1 is accelerated in free fall, until it has reached velocity v. An external force acts upon clock U2, preventing it from being set in motion by the gravitational field. When the clock U1 has reached velocity v the gravitational field disappears.
2. U2 moves with constant velocity v up to point B of the positive x-axis. U1 remains at rest.	2. U1 moves with constant velocity to the point B' on the negative x-axis. U2 remains at rest.
3. Clock U2 is accelerated by an external force acting in the negative direction of the x-axis until it has reached velocity v in the negative x-direction. U1 remains at rest.	3. A homogenous gravitational field appears, that is directed towards the positive x-axis. Clock U1 is accelerated in the direction of the positive x-axis until it has reached the velocity v, then the gravitational field disappears again. An external force, acting upon U2 in the negative direction of the x-axis prevents U2 from being set in motion by the gravitational field.
4. U2 moves with constant velocity in the negative direction of the x-axis until it approaches U1. U1 remains at rest.	4. U1 moves with constant velocity v in the direction of the positive x-axis until it approaches U2. U2 remains at rest.

K is the inertial reference frame.	K' is the accelerated reference frame.
5. An external force brings clock U2 to rest.	5. A gravitational field that is directed towards the negative x-axis appears and brings U1 to a halt. Then the gravitational field disappears again. An external force keeps U2 in a state of rest.

It should be kept in mind that in the left and in the right section exactly the same proceedings are described, it is just that the description on the left relates to the coordinate system K, the description on the right relates to the coordinate system K'. According to both descriptions, the clock U2 is running a certain amount behind clock U1 at the end of the observed process. When relating to the coordinate system K' the behavior explains itself as follows: During the partial processes 2 and 4 the clock U1, going at a velocity v, runs indeed at a slower pace than the resting clock U2. However, this is more than compensated by a faster pace of U1 during partial process 3. According to the general theory of relativity, a clock will go faster the higher the gravitational potential of the location where it is located and during partial process 3 U2 happens to be located at a higher gravitational potential than U1. The calculation shows that this speeding ahead constitutes exactly twice as much as the lagging behind during the partial processes 2 and 4. This consideration completely clears up the paradox that you brought up.

From the point of view of the stay on the earth, the age difference is due to the fact that K' is accelerated, but from the point of view of the traveler, this age difference comes from the fact that K is in a gravitational field. The Einstein analysis is important because it gives the scheme for a precise calculation of the times measured by the two observers, as we shall see in Chapter 6.

However, the critic is not completely satisfied. He asks whether the gravitational field that Einstein introduced was not a fictitious field. Einstein first shows that the concepts "real" and "fictitious" are not so simple that a detailed analysis is necessary. In addition, he gives a long answer based on the concepts of the General Theory of Relativity. One of the reasons not to call this field "fictitious" is given by Einstein:

> it is not an *a priori* necessity that the particular concept of the Newtonian theory, according to which every gravitational field is conceived as being brought forth by mass, should be retained in the general theory of relativity.

We have a very clear analysis of the travel to the star from two points of view: that of the stay on earth and that of the traveler. This analysis suggests that for the stay on earth, one can apply the Special Theory of Relativity, but for the traveler, the General Theory is necessary. We show later that there was a long discussion as to whether the acceleration is an indispensable ingredient of the solution. From the analysis of Einstein himself, there is no place for such a discussion and we show in another chapter the consequences of this discussion.

The Einstein solution elucidates the qualitative explanation of the Clock paradox. The traveler is younger than the stay at home because the traveler is accelerated relative to the stay at home. For the traveler, the stay at home is older because the latter is in a gravitational field. More exactly, the traveler (in his own frame) sees the stay at home moving in an accelerated motion. He knows that the stay at home does not have an engine that could explain this accelerated motion. Hence, the idea that the stay at home is in a gravitational field. Each observer sees the other observer moving in an accelerated motion but with different consequences: the stay at home is older than the traveler, and the traveler is younger than the stay at home.

5.5. The Born Solution

We present here a solution published by Born in 1922, by Kopff in 1923 and by Tolman in 1934. This solution was reproduced several times in publications which are more recent. From the point of view of the stay at home, things are very simple. Since the time of application of the acceleration is very short, the distances at which the acceleration is applied are very small and negligible relative to the earth–star distance L_0. For the stay on earth, the travel time ΔT is merely $\Delta T = 2L_0/V$, where V is the absolute value of the velocity. We can calculate the time of the traveler as an application of the time dilation and the traveler's time is

$$\Delta\tau = \frac{2L_0}{V}\sqrt{1 - \frac{V^2}{c^2}} = \frac{2L}{V}, \tag{5.12}$$

with $L = L_0\sqrt{1 - \frac{V^2}{c^2}}$. In (5.12), it is supposed that at the stopping, the clock of the traveler shows a smaller value than that of the clock of the earth. Without the stopping, the two observers are in a symmetrical position and there is no age difference (see Chapter 3). The stopping is an important ingredient of this model even though it is not always mentioned.

Now, one considers the point of view of the traveler. During the short times when an acceleration is applied at the turnover, this is equivalent to gravitational field. During the start and the landing, the two observers are in the same frame at zero distance and there is no gravitational field. Only in the reverse, when the observers are at the distance L_0 to one another, there is a potential equal to the product acceleration distance or $\Phi = gL_0$. The total time of the stay at home is the sum of the gravitational effect and the time dilation. The first effect takes place during a short time Δt and since the change in the velocity is $2V$, the acceleration is $g_0 = 2V/\Delta t$. The relation between the time Δt and the change in the time of the stay on earth is as follows (see expression (4.27)):

$$\Delta'T = \left(1 + \frac{L_0 g_0}{c^2}\right)\Delta t = \left(1 + \frac{2VL_0}{c^2\Delta t}\right)\Delta t, \tag{5.13}$$

$$\Delta'T = 2V/g_0 + \frac{2VL_0}{c^2}, \tag{5.14}$$

taking into account that $\Delta t = 2V/g_0$. If the acceleration g_0 is very large, $\Delta'T$ is reduced to $\frac{2VL_0}{c^2}$ and this can be interpreted as a discontinuity in the time of the stay at home from the point of view of the traveler. We recall that we already met this discontinuity in the Langevin model above.

The time dilation gives

$$\Delta''T = \left(\frac{2L}{V}\right)\sqrt{1 - \frac{V^2}{c^2}} = \frac{2L_0}{V}\left(1 - \frac{V^2}{c^2}\right). \qquad (5.15)$$

The total time of stay on earth is

$$\Delta T = \Delta'T + \Delta''T = \frac{2L_0}{V} + 2V/g_0. \qquad (5.16)$$

If now g_0 is very large, one finds merely that the time of the stay on the earth is $2L_0/V$. In conclusion, the two observers agree that the travel time of the stay on the earth is $2L_0/V$ and that of the traveler is $\frac{2L_0}{V}\sqrt{1 - \frac{V^2}{c^2}}$. The ratio of the proper times of the two frames, $\Delta\tau/\Delta T$, is equal to γ and it is characteristic of all the models with negligible acceleration times.

One remarks that the acceleration g_0 cannot go to large values without keeping the product $g_0\Delta t$ constant since it is equal to $2V$.

In this solution, we note two points of view. For the stay on earth, his age is that of the travel time L_0/V. From the point of view of the traveler, the travel time of the stay on earth is the sum of the time dilation that is applied by the traveler (in the whole solution, it is applied twice) and that of the gravitational field.

In this solution, the role of the acceleration is seen at the turnover. But what would happen if the traveler remains on the star and does not come back immediately? We can imagine that the clocks on the earth and on the star are synchronous such that when the traveler lands on the star, he can see directly what is the proper time on the earth. It is not difficult to adapt the preceding calculation to the case where the traveler remains on the star (Cochran, 1960). The clocks (that of the traveler and that of the star) will indicate $L_0/(\gamma V)$ and L_0/V respectively.

This result is very important because this gives the distinction between the two types of solutions: those based only on the Special Theory of Relativity and those using the principle of equivalence. In the first type, the age difference appears at the turnover when the traveler changes his direction (Langevin solution). In the second type, the turnover has no role: the traveler may remain on the star for the time he wants; the clocks of the two observers indicate already different times.

One may wonder about the viewpoint of Langevin. In his 1911 article, the acceleration is given as the basis for the age difference, but in his solution of 1922, it is presented as the consequence of the turnover. One cannot conclude that Langevin had changed his mind, because in 1923, he published a book *La physique depuis vingt ans* in which he included his 1911 article. To solve this apparent contradiction, one can say that Langevin sees in the turnover the consequence of the acceleration and that is his interpretation of how the acceleration works. This also means that he does not relate the acceleration to the principle of equivalence. However, in a final remark, Langevin makes an allusion to the gravitational field and this means that he knew the principle of equivalence, but did not use it.

Chapter 6

The Einstein–Møller Solution

6.1. Determination of the Travel Times

It is really astonishing to note that after the Einstein paper of 1918 quoted in Chapter 5, it was only in 1952 that Møller published a solution for the thought experiment of Langevin following the scheme of Einstein. We follow closely the calculations of Møller's book.

We now recall the scenario of the experiment. The journey of the traveler toward the star is composed of three parts (Fig. 6.1). In the first part, between the points E (the earth) and B, a constant force F is applied to the traveler until he reaches velocity V at the point B. In the second part, he goes with constant velocity V between the points B and C. Finally, at C, the same force F is applied again but in the opposite direction to the motion in order to decrease the velocity. At the point S (the star), the velocity becomes null and the motion is reversed, both under the influence of the force $-F$.

The return travel is symmetric to the first path. Along SC the force $-F$ is applied in the direction of the motion, which is now accelerated. At C, the traveler reaches again the velocity V (and by symmetry, one has $EB = CS$) and finally, at B, the force F is applied in the direction $x > 0$ in order to decrease the velocity to zero at the point A. It is obvious that the motions in the four portions EB, CS, SC, and BE are identical to each other. This is the scenario from the point of view of the stay on the earth (Fig. 6.1).

For the traveler, things are different. Figure 6.2 shows the different steps of the first part of the journey, as seen from the point of view of the traveler who is supposed to be at rest in his frame.

87

$$ES = L \qquad\qquad EB = CS = X_1$$

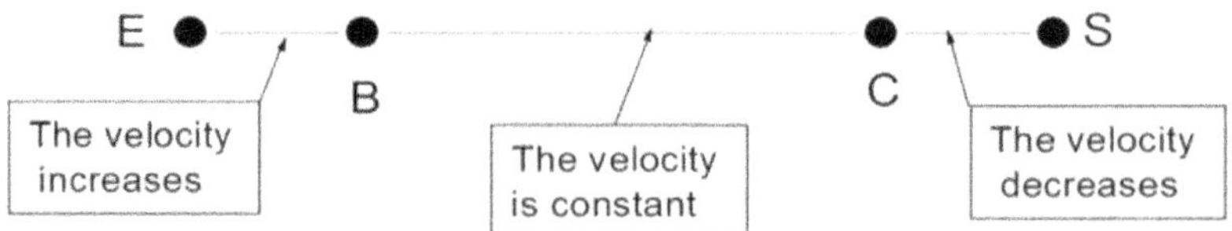

Fig. 6.1. The travel from the earth, E, to the star, S, in three parts from the point of view of the stay on the earth. EB, velocity increases under the influence of the force F; BC, velocity is constant ($F = 0$); CS, under the force $-F$, velocity decreases until it becomes null. The return travel corresponds to: SC, velocity increases; CB, velocity is constant; BE, velocity decreases.

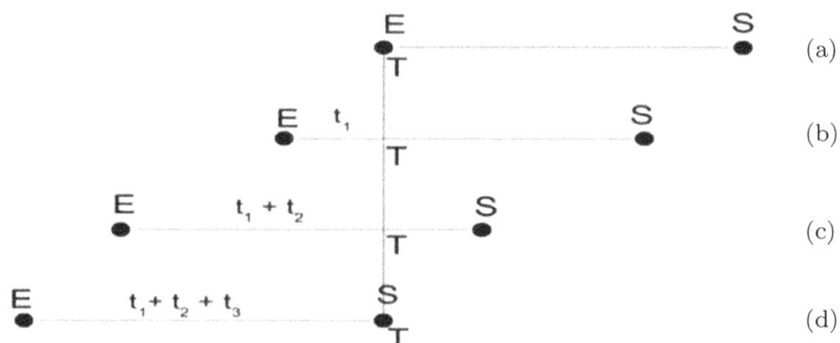

Fig. 6.2. Positions of the earth and the star from the point of view of the traveler T: (a) at the time $t = 0$; (b) at the end of the time where the traveler (in the earth frame) is accelerated and the stay on the earth is under a gravitational field; (c) at the end of the time where the stay on the earth has moved with constant velocity (in the frame of the earth); (d) the star has reached the traveler.

Figure 6.2(a) corresponds to the starting time. Until the time t_1, the earth is falling freely in the negative x direction under the influence of the gravitational field (Fig. 6.2(b)). Figure 6.2(c) describes the state after the time $t_1 + t_2$ while between t_1 and t_2 the earth moves with a constant velocity V. Finally, the star S reaches the traveler, and the earth is now at the longest distance. The time elapsed from the beginning is $t_1 + t_2 + t_3$ while between t_3 and t_2 there is a gravitational field.

We shall consider separately the points of view of the stay on the earth in his frame S_1 and the traveler in his frame S_2.

6.1.1. *The point of view of the observer on the earth*

One can easily determine the proper times of the two observers, as measured or calculated by the stay at home from the motion equation (4.13) given in Chapter 4. It suffices to calculate the travel time from the earth to the star, since the return travel takes the same time.

The equation of the motion is as indicated:

$$\frac{d}{dT}\left(\frac{mv}{\sqrt{1-\frac{v^2}{c^2}}}\right) = F.$$

From this equation, one obtains the velocity v by an integration

$$v = \frac{gT}{\sqrt{\left(1+\left(\frac{gT}{c}\right)^2\right)}}. \tag{6.1}$$

The parameter g is equal to F/m. It has the same dimensions as but is not acceleration. Only for small velocities is the ratio F/m equal to the acceleration g_0. By a second integration, one obtains the distance X as a function of the time T.

$$X(T) = \frac{c^2}{g}\left[\sqrt{1+\left(\frac{gT}{c}\right)^2} - 1\right]. \tag{6.2}$$

After a time $\Delta'T$, the traveler reaches the point B where the velocity becomes V. One has for the distance $EB = X_1$ and the time $\Delta'T$ from (6.1) and (6.2) as functions of V through the groups $\beta = V/c$ and $\gamma = \frac{1}{\sqrt{1-\beta^2}}$

$$X_1 = \frac{c^2}{g}(\gamma - 1), \tag{6.3}$$

$$\Delta'T = \frac{c}{g}\gamma\beta. \tag{6.4}$$

The distance BC is $L - 2X_1$ and the time $\Delta''T$ taken to travel it is $(L - 2X_1)/V$ or

$$\Delta''T = \frac{L}{V} - \frac{2c(\gamma - 1)}{g\beta}. \tag{6.5}$$

The travel time of the distance CS is equal to the travel time of EA by symmetry and one obtains the total time of the travel, from the point of view of the stay on the earth, as

$$\Delta T = 2\Delta'T + \Delta''T = \frac{L}{V} + \frac{2c(\gamma - 1)}{g\beta\gamma}. \tag{6.6}$$

Now, the observer on the earth can calculate the travel time as measured by the traveler, i.e., the proper time of the traveler, by the use of equation which takes the following form:

$$\Delta\tau = \int_0^{\Delta T} dT \sqrt{1 - \frac{V(T)^2}{c^2}},$$

where $\Delta\tau$ is the proper time of the traveler between the earth and the star. As to the stay on the earth, the time between E and B is equal to the time between C and S. The expression above becomes

$$\Delta\tau = 2\Delta'\tau + \Delta''\tau = 2\int_0^{\Delta'T} dT \sqrt{1 - \frac{V(T)^2}{c^2}} + \Delta''T/\gamma. \tag{6.7}$$

The last term in the right-hand side of (6.7) corresponds to the time dilation as seen by the traveler for the travel between B and C. $V(T)$ is given by (6.1) and one obtains

$$1 - \frac{V^2}{c^2} = \frac{1}{1 + \frac{(gT)^2}{c^2}}, \tag{6.8}$$

$$\Delta'\tau = \int_0^{\Delta'T} dT \frac{1}{\sqrt{1 + \frac{(gT)^2}{c^2}}} = \frac{c}{g}\sinh^{-1}\left(\frac{g\Delta'T}{c}\right) \tag{6.9}$$

or

$$g\Delta'T/c = \sinh\left(\frac{g\Delta'\tau}{c}\right). \tag{6.10}$$

We use the identity

$$\tanh(x) = \frac{\sinh(x)}{\cosh(x)} = \frac{\sinh(x)}{\sqrt{1 + \sinh(x)^2}} \tag{6.11}$$

in order to write from (6.10)

$$\tanh(g\Delta'\tau/c) = \frac{g\Delta'T/c}{\sqrt{1 + (g\Delta'T/c)^2}}. \tag{6.12}$$

Taking $\Delta'T = \frac{c}{g}\gamma\beta$, one obtains

$$\tanh\left(\frac{g\Delta'\tau}{c}\right) = \beta \tag{6.13}$$

and finally

$$\Delta'\tau = \frac{c}{g}\tanh^{-1}(\beta). \tag{6.14}$$

Now, we can calculate the proper time of the traveler from (6.7) and $\Delta''T$ given by (6.5)

$$\Delta\tau = \frac{L}{V\gamma} + \frac{2c}{g}\left[\tanh^{-1}(\beta) - \frac{(\gamma - 1)}{\beta\gamma}\right]. \tag{6.15}$$

We have now the times of the two observers and we have to verify that the time of the traveler is shorter than the time of the stay on the earth or $\Delta\tau < \Delta T$. This gives the following inequality:

$$\frac{Lg}{2c^2} > \beta\frac{\gamma}{(\gamma - 1)}\tanh^{-1}(\beta) - 2. \tag{6.16}$$

It is not possible to verify directly this inequality, since the velocity V and g are related through $\Delta'T$ (see equation (6.1)).

We propose to use the following method. In the problem, there is one characteristic time $T_L = L/c$; it is the time required by light to go from the earth to the star in the frame of the earth. It is obvious that the time $\Delta'T$ of the first part of the travel is shorter than T_L. One supposes that

$$\Delta'T = \alpha T_L \tag{6.17}$$

with $\alpha < 1$, and equation (6.17) can be written as

$$\frac{c\beta\gamma}{g} = \alpha\frac{L}{c} \tag{6.18}$$

or

$$\frac{Lg}{2c^2} = \frac{\beta\gamma}{\alpha} > \beta\frac{\gamma}{(\gamma - 1)}\tanh^{-1}(\beta) - 2. \tag{6.19}$$

Since $1/\alpha > 1$, it suffices to verify that

$$\beta\gamma > \beta\frac{\gamma}{(\gamma - 1)}\tanh^{-1}(\beta) - 2 = f(\beta) \tag{6.20}$$

in order that (6.19) be verified. It is not difficult to show that inequality (6.20) is satisfied for all values of β and γ (i.e., the velocity V; in the first part of the travel at constant velocity). However, before concluding that the traveler's time is always shorter than that of the stay on the earth, one has to check another inequality.

The distance $EA = X_1$, in the first part of the travel, cannot be longer than $L/2$. Taking the expression of X_1 given by $[c^2/g](\gamma - 1)$, this gives

$$\frac{c^2}{gL} < \frac{1}{2(\gamma - 1)}. \tag{6.21}$$

The meaning of (6.21) is that, for a choice g of the power engine (since $g = F/m$), there is a limitation of γ, i.e., V. However, in all cases, the age of the traveler is smaller than that of the stay on the earth. We discuss below this point in detail.

At this stage, one can say that the problem of Langevin is solved, the theory gives the correct answer, and Langevin is right. However, it will be very interesting to understand the point of view of the traveler following the lines of Einstein.

6.1.2. *The point of view of the traveler*

Now one supposes the frame S_2 at rest and one calculates the travel time for the stay at home observer. In the time intervals where S_2 is accelerated relative to S_1, there is a gravitational field in S_2. The gravitational potential is given by $\Phi = -gx(1 - gx/2c^2)$, where x is the distance between the two frames. We recall that from the Theory of General Relativity, two clocks separated by a distance x in the direction of a gravitational field have different rates such that the times are related by $t_2 = t_1[1 + \Phi(x)/c^2]$, where the clock at the lower potential has the smaller rate. This is the physical basis for the determination of the travel time in the frame S_2. Møller, using the tools of General Relativity, made the calculation.

Møller divides the travel time of the stay at home as seen by the traveler into three parts: the initial time of the application of the force ΔT_1, the time at constant velocity ΔT_2, and finally, the time of stopping until arrival at the star ΔT_3.

The final result is that the elapsed time $(\Delta T = \Delta T_1 + \Delta T_2 + \Delta T_3)$ on the earth as measured in the frame of the traveler is given by

$$(\Delta T) = L/V + (2c/g)(\gamma - 1)/\gamma\beta,$$

which is identical to the time elapsed on the earth as measured by the stay at home observer. In other words, the proper times measured in both frames S_1 and S_2 are the same. This result was expected. Since the traveler and the stay at home are in the same reference frame at the beginning and at the end, one finds the same age for each observer whatever the method for determination.

6.2. Analysis of the Travel Times

We shall compare the two proper travel times of the two observers under different conditions. We now recall the following:

- Proper time of the stay on the earth, given by (6.6):

$$\Delta T = \frac{L}{V} + \frac{2c(\gamma - 1)}{g\beta\gamma}.$$

- Proper time of the traveler in (6.15):

$$\Delta\tau = \frac{L}{V\gamma} + \frac{2c}{g}\left[\tanh^{-1}(\beta) - \frac{(\gamma-1)}{\beta\gamma}\right].$$

It is possible to write again the proper times as follows:

$$\Delta T = \frac{L}{c}\left[\frac{c}{V} + \frac{2c^2}{gL}\left(\frac{\gamma-1}{\beta\gamma}\right)\right]. \tag{6.22}$$

The group $\frac{c^2}{gL}$ can be seen as the ratio of two characteristic times of the problem. We already saw that $\frac{L}{c} = T_L$. The second characteristic time is given by $\frac{c}{g} = T_A$. So one can write

$$\frac{c^2}{gL} = \frac{T_A}{T_L} \tag{6.23}$$

and

$$\frac{\Delta T}{L/c} = \delta T = \frac{1}{\beta} + \frac{2T_A}{T_L}\left(\frac{\gamma-1}{\beta\gamma}\right). \tag{6.24}$$

One can also write the proper time of the traveler normalized relative to (L/c) and one obtains

$$\frac{\Delta\tau}{L/c} = \delta\tau = \frac{1}{\beta\gamma} + \frac{2T_A}{T_L}\left[\tanh^{-1}(\beta)\frac{\gamma-1}{\beta\gamma}\right]. \tag{6.25}$$

These two expressions (6.34) and (6.25) have meaning only if inequality (6.21) is satisfied. It can be written as

$$\frac{T_A}{T_L} < \frac{1}{2(\gamma-1)}. \tag{6.26}$$

From (6.26), one can obtain the largest possible velocity for a given choice of the ratio T_A/T_L. This is shown in Fig. 6.5.

As a final step, one can compare the results of this chapter with the results of Chapter 5 about the Born solution. We recall the results but change the exact formulation, because:

(1) the calculation in the present chapter concerns only the earth–star travel;

(2) in the present chapter, the earth–star distance is denoted by L and not L_0.

The time of the stay on the earth (from E to S) is given by the Born solution:

$$\Delta T = \frac{L}{V} + \frac{V}{g} \tag{6.27}$$

and by the Møller solution as (6.6):

$$\Delta T = \frac{L}{V} + \frac{2c(\gamma - 1)}{g\beta\gamma}.$$

It is not difficult to show that expression (6.6) goes to (6.27) if the ratio $\beta = V/c$ is small. If β is small, γ can be written as $\gamma \cong 1 + \frac{\beta^2}{2}$, and inserting into (6.6), one obtains (6.27). Similarly, one can show that the expression (6.13) for $\Delta\tau$ (traveler time) is reduced to $\frac{L}{V}\sqrt{1 - \frac{V^2}{c^2}}$ for small velocities. This means that the Born solution is good only for large values of g and relatively small velocities. The ratio T_A/T_l is dependent on two parameters of the problem: the quantity $g = F/M$ and the earth–star distance. If this ratio is small, the traveler can choose a large velocity. This is possible either if his engine is powerful enough (g is large) and/or if the star is far from the earth.

From a physical point of view, the difference between the two solutions (Born and Møller) lies in the different expressions of the gravitational field. For Born, it is a delta function, and for Møller, it is a quadratic function dependent on the distance between the observers.

We shall now consider some specific cases. The first case is given by $T_A/T_l \to 0$ or $g \to \infty$ (very short time of acceleration). Consequently, from equations (6.6) and (6.15), one obtains $\Delta T = L/V$ and $\Delta\tau = L/(V\gamma)$. In Fig. 6.3, the reduced times δT and $\delta\tau$ are given as functions of the reduced velocity that the traveler may choose.

One has to note that the condition $g \to \infty$ is associated with another condition: the product $g\Delta'\tau$ must go to a finite limit. This

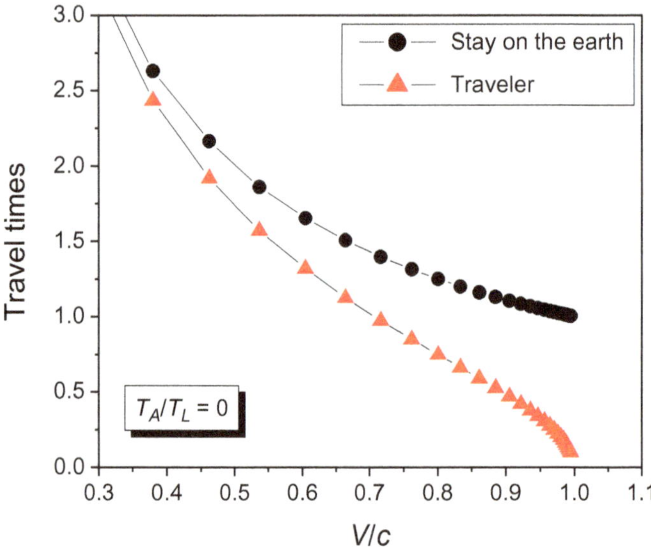

Fig. 6.3. Reduced travel times for the stay on the earth and for the traveler when $g \to \infty$. In this case, the ratio of the two times is the largest and equal to γ.

can be seen from equation (6.13)

$$\tanh\left(\frac{g\Delta'\tau}{c}\right) = \beta.$$

It is not possible to do $g \to \infty$ alone, since in such a case, $[\tanh(\frac{g\Delta'\tau}{c})]$, goes to 1, which is not possible because β is smaller than 1. Consequently, the product $[g\Delta'\tau]$ must be finite. One recalls that $\Delta'\tau$ is the time of the first part or the third part of the travel (force different from zero) in the traveler's frame. For small velocities, (6.13) becomes $\Delta'\tau \cong V/g$.

However, Møller gives a surprising result. One expects that the two times ΔT_1 and ΔT_3 of application of the force will go to zero when the central part is equal to L/V. The three times that give the travel time are

$$\Delta T_1 = V/g, \quad \Delta T_2 = \Delta T''\left(1 - \frac{V^2}{c^2}\right), \quad \Delta T_3 = \left(\frac{V}{g} + \frac{L}{c}\right)\left(\frac{V}{c}\right)$$

with $\Delta T''$ given by equation (6.5)

$$\Delta T'' = \frac{L}{V} - \frac{2c(\gamma - 1)}{g\beta}.$$

We recall that the sum $\Delta T_1 + \Delta T_2 + \Delta T_3$ gives the stay at home's proper time, equation (6.6). However, when one considers the limits of the three times ΔT_i for $g \to \infty$, one sees that the time ΔT_1 goes to zero as expected, the time ΔT_2 goes to $L/\gamma V$, but ΔT_3 does not go to zero but to LV/c^2. In other words, at the turnover, the time of the stay at home shows a discontinuity. This peculiarity is due to the fact that the gravitational potential is infinite at this point. In the present case, the discontinuity is only LV/c^2 since one considers only the earth–star travel. For the complete journey, this discontinuity is $2LV/c^2$. We have already met twice this discontinuity: first in the Langevin model and secondly in the Born model. We stress again that this unrealistic discontinuity was the consequence of the choice $g \to \infty$.

Now, one considers the case when the ratio T_A/T_L is different from 0: it is taken equal to 0.5. The reduced travel times as a function of β are given in Fig. 6.4. One sees that the velocity is limited and the traveler cannot travel faster than about 0.86c. Physically, this is explained by the fact that he can reach only the half distance to the star at this maximum velocity compatible with the power of his engine, and so he is obliged to reduce the velocity in order to land on the star.

Finally, we shall give some details of this situation without time dilation. The traveler reaches the half distance without stopping his engine and his velocity increases regularly. In the second part (after passing the half distance), he reverses the direction of the force until he stops on the star. The inequality (6.32) is now an equality

$$\frac{T_A}{T_L} = \frac{1}{2(\gamma - 1)}. \tag{6.28}$$

From the expressions for δT (6.20) and $\delta \tau$ (6.21), one can obtain the travel time ratio as a function of the T_A/T_L as well as the maximum velocity (Fig. 6.5).

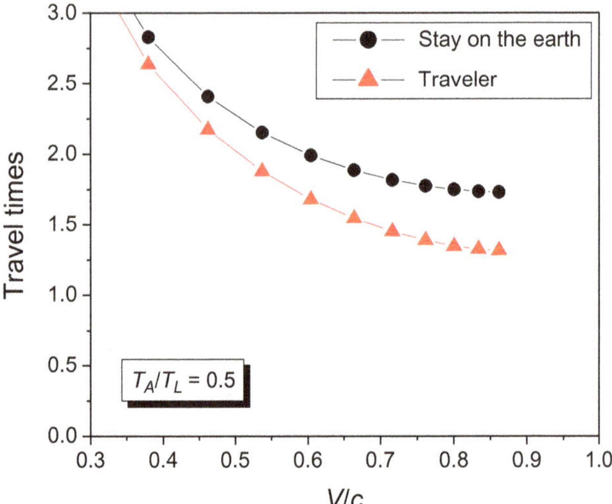

Fig. 6.4. Reduced travel times for $T_A/T_l = 0.6$. The traveler cannot choose a velocity larger than about $0.86c$.

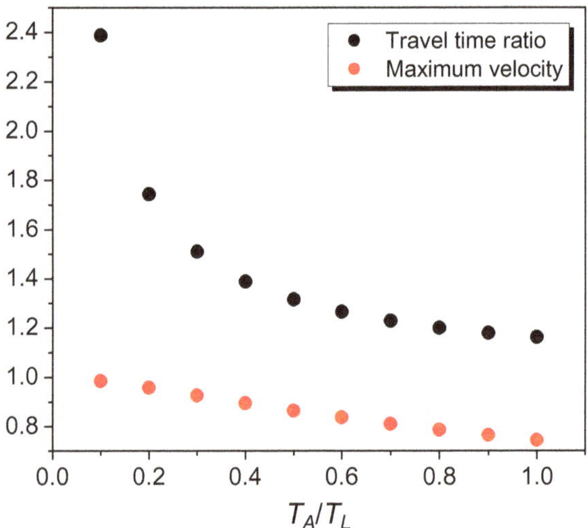

Fig. 6.5. Ratio of the travel times in the case where there is no time dilation. The maximum velocity is reached at the half distance to the star.

This last case with no time dilatation was developed by Marder (1971) and by Iorio (2005). These articles are very important since they demonstrate that acceleration is an important ingredient for understanding the Clock paradox. There are so many articles in which the authors consider only the time dilation as the sole mechanism to understand the age difference.

To conclude, some remarks will be presented concerning the qualitative behavior of the travel. From the point of view of the stay on the earth, the travel time is longer than L/V, since in the first part and the third part, the mean velocity is smaller than V. The travel time of the traveler is smaller than that of the stay on the earth for two reasons: the time dilation and acceleration. The two effects can appear independently or together. In particular, even if the second part of the travel (in which the velocity of the traveler is constant) shrinks to zero and there is no time dilation, the traveler is always younger. From the point of view of the traveler, the longer travel time of the stay on the earth is due to the gravitational field. We recover a qualitative interpretation of the Langevin experiment and we understand qualitatively the age difference of the two observers.

In his book, Møller gives a solution to the circular Twin paradox. The traveler does not do a round trip to a star but travels along a circle. He begins at some point of the circle where the stay at home is located and come back to this point. As expected, the traveler's time is shorter than the time of the stay at home. This is the classical problem of the rotating disk in Special Relativity, which was used to expose the principle of equivalence. It is possible to show that the following equation relates the time of the traveler t to the stay at home's time t_0

$$t = t_0 \sqrt{1 - \frac{(\omega r)^2}{c^2}}, \qquad (6.29)$$

where ω is the angular velocity of the disk and r its radius. The centrifuge potential can be seen as a gravitational potential Φ such that $\Phi = -\frac{(\omega r)^2}{2}$ and (6.29) can be written as

$$t = t_0 \sqrt{\left(1 + \frac{2\Phi}{c^2}\right)}. \qquad (6.30)$$

This expression is not identical to equation (4.25). But we saw that (4.25) is only an approximation while (6.30) is the correct one. It can be easily seen that (6.30) reduces to (6.25) for small potentials.

6.3. The Hafele–Keating Experiment

We describe briefly the Hafele–Keating experiment, which may be seen as a special version of the rotating Clock paradox. One considers two clocks: one is at rest at some position on the earth at the sea level. The second clock goes around along an equatorial line with radius R and velocity V. A question arises: relative to what reference frame is this velocity defined. Hafele and Keating take a hypothetical inertial frame in which the earth rotates. It results that V is equal to $R\omega + p$, where ω is the angular velocity of the earth and p the velocity of the clock relative to the earth. Note that p can be negative or positive depending on the direction of the travel clock.

The relativistic relation of time dilation applied to the mobile clock at the sea level is

$$\tau_1 = \tau_0 \sqrt{\left(1 - \frac{V^2}{c^2}\right)} = \tau_0 \left(1 - \frac{(R\omega + p)^2}{2c^2}\right) \tag{6.31}$$

since $V = (R\omega + p) \ll c$. Here $\tau_0 = 2\pi R / |p|$ is the time measured by the clock in the inertial frame during the travel around the earth. The clock on the ground is also moving in the inertial frame and its time τ_2 is related to τ_0 by

$$\tau_2 = \tau_0 \sqrt{\left(1 - \frac{(R\omega)^2}{c^2}\right)} \cong \tau_0 \left(1 - \frac{(R\omega)^2}{2c^2}\right). \tag{6.32}$$

The time difference between the mobile clock and the clock on the earth is

$$\tau_1 - \tau_2 = -\left(\frac{2pR\omega + p^2}{2c^2}\right)\tau_0 = -\left(\frac{2pR\omega + p^2}{2c^2}\right)\frac{2\pi R}{|p|}. \tag{6.33}$$

But there is another source of time difference: it is the fact that the mobile clock is located at h meters above the sea level and there is a gravitational potential difference Φ between the two clocks. As was seen in Chapter 4, their time difference is given by

$$\tau_3 = \tau_2 \left(1 + \frac{\Phi}{c^2} \right) = \tau_2 \left(1 + \frac{gh}{c^2} \right), \tag{6.34}$$

where g is the gravitation constant on the earth. From (6.34), one obtains

$$\tau_3 - \tau_2 = \frac{gh}{c^2} \tau_2 = \frac{gh}{c^2} \left(1 - \frac{R^2 \omega^2}{2c^2} \right) \tau_0 = \frac{gh}{c^2} \frac{2\pi R}{|p|} \tag{6.35}$$

since $\frac{R^2 \omega^2}{2c^2} \ll 1$.

The total time difference is

$$\Delta\tau = \left(\frac{gh}{c^2} - \left(\frac{2pR\omega + p^2}{2c^2} \right) \right) \frac{2\pi R}{|p|}. \tag{6.36}$$

One has two different expressions depending on the sign of p, i.e., the direction of the mobile clock.

- Eastward, p is positive and equation (6.36) becomes

$$\Delta\tau_1 = \frac{2\pi R}{c^2} \left(\frac{gh}{p} - R\omega - \frac{p}{2} \right). \tag{6.37}$$

- Westward, p is negative and one writes $p = -a$ ($a > 0$) and equation (6.36) becomes

$$\Delta\tau_2 = \frac{2\pi R}{c^2} \left(\frac{gh}{a} + R\omega - \frac{a}{2} \right). \tag{6.38}$$

Comparing (6.37) and (6.38), one sees that $\Delta\tau_1$ is smaller than $\Delta\tau_2$ as it was found in the experiment of Hafele and Keating. To calculate the exact values of the times, it is necessary to know the height h and the velocity p. Hafele and Keating give $\Delta\tau_1 = -59 \pm 10$ and $\Delta\tau_2 = 273 \pm 7$ (nanoseconds).

6.4. The Precursor and the Followers

Since the publication of the book by Møller, one can say that there are two kinds of publications in the literature on the Clock paradox: first, those which reduce the time of application of acceleration such that it is negligible and, secondly, those which take into account acceleration explicitly. We note that the number of publications of the first kind is much larger than the second kind. Here we wish to only mention some publications in which acceleration is taken into account completely. Probably, the precursor of the second group is Campbell, who published an article in 1933 and showed that whatever the path of the traveler, if he comes back to the position of the stay at home, the traveler's time is always smaller than the time of the stay at home. The regular Twin paradox and the circular Twin paradox are particular cases of the general proof of Campbell.

Finally, some authors present a complete solution, often based on the Møller solution (Special Relativity and General Relativity). The most remarkable article is that of Perrin, who complains about the paucity of articles with a complete calculation of the times from the point of view of the traveler. The articles of Leffert and Donahue (1958), Good (1981), Müller *et al.* (2008), Minguzzi (2004) and Wu and Lee (1972) propose different solutions for the complete calculation of the times of the two observers. It is likely that there are also other publications.

Chapter 7

The Disappearance of the Acceleration and the Symmetry Breaking

Very soon after the publication of the Langevin article, there were two debates, which are not completely independent. Langevin introduces a non-inertial frame (that of the traveler), and it seems at first glance that the General Relativity is the most adapted method since that Special Theory of Relativity deals only with inertial frames. At the time of publication of the Langevin article (1911), Einstein had not published his General Theory of Relativity and one may have the impression that Langevin proposed a problem without possibility of solution. Thus the question arises: what are the tools needed to solve the paradox? Are they present in the Special Relativity? Or maybe the General Theory of Relativity is needed? These questions were at the center of a debate about the choice of the theory: Special or General Relativity. Meanwhile, a second debate took place by asking a surprising question: is acceleration a necessary ingredient or not? Some authors answered that acceleration is not needed but is useful only for setting the traveler in motion.

It is clear that the tendency to eliminate acceleration is related to the belief that in Special Relativity, there is no place for acceleration. Consequently, it is proposed that in order to solve the Langevin problem in the frame of Special Relativity, acceleration must be put aside. Sometimes there is no mention at all of acceleration. In those last cases, one does not know how the traveler begins his motion nor how he comes back!

Certainly, those asking this question had not read Langevin's article.

7.1. How to Clear the Acceleration?

The efforts to minimize the role of acceleration can be classified into two possibilities. The first tries to eliminate the acceleration but retaining it: one takes very short acceleration times and goes on as if acceleration is non existent. The second possibility is to eliminate the acceleration completely, creating a new model, the three-frame model. In spite of the absence of acceleration, it is reported (Grünbaum, 1954) that this model is equivalent to the original problem of Langevin.

In the first solution, acceleration has only one benefit. It permits to give immediately to the traveler the velocity he needs. Nevertheless, there is a question (we shall see later that it induced discussion): where is the acceleration? It does not appear in the expression of the travel time and this seems strange. Arzeliès (1966) tries to answer this but he is not convincing. In other words, there is a need for an acceleration (since without it there is no travel), but the solution of the problem is independent of the acceleration. This is a troubling situation and in fact, very few are aware of this "paradox".

If one eliminates acceleration, one has to ask immediately: how is such a situation possible? The traveler takes off, decelerates when approaching the star and stops on it. At each step, there is acceleration. Some people ignoring the acceleration are not bothered by this question, since very often they consider only the point of view of the stay on the earth who is never in an accelerated frame. The habitual answer to this question is to adopt a three-frame model (explicitly or very often implicitly) as did many scientists.

As its name indicates, in the model without acceleration, there are three different frames. The first frame is that of the stay at home (frame A), the second frame is that of a traveler (frame B), and the third frame C is that of another traveler. The traveler B comes from infinity ($x < 0$) with the velocity V and crosses, at the time $t = 0$, the frame A. He goes on to the star (he does not stop) and continues to infinity, but exactly when he reaches the star, he crosses a third traveler (frame C) who travels toward the frame A with the velocity $-V$. At the crossing, travelers B and C synchronize their clocks. When the traveler C crosses A, they read the times indicated by their respective clocks. C goes to infinity as B does. In principle, the travelers B and C mimic the unique traveler of Langevin. The first idea was to transfer the traveler from his original frame to another frame traveling in the negative direction after reaching the star. But in order to avoid a terrible trauma, the three-frame model was proposed to the traveler.

This proposed model is analogous to the Langevin experiment, but it is really different for this simple and unique reason: in the three-frame model, the observers are never in the same frame and even if they can know their own proper times, they cannot know directly the proper times of the other observers. Consequently, they cannot compare their ages directly. For example, at the end of the story when A and C meet (at the crossing but not in the same frame!), A reads the time of C differently from the time read by C himself. Consequently, the comparison of the readings is not the comparison of their proper times, i.e., their ages.

A second point is more astonishing. During the first part of the travel (B leaves A) the observers A and B are in symmetrical positions. This means that the travel times are equal to L/V for both observers A and B. And in the second part, one has again symmetry between A and C and the travel time is for both observers L/V. As a result, the observers A and C are exactly the same age at their crossing and there is no age difference!

It is often repeated that this three-frame model is equivalent to the original model of Langevin and the conclusion is that there is no necessity to introduce acceleration to solve the problem.

It is surprising to read that the Einstein paper of 1918 (the debate with Criticus) is illogical, since it supposes that the acceleration is essential for the solution. The author (Unnikrishnan, 2005) reaches this conclusion, since for him, the three-frame model is the counterexample to the assertion of Einstein!

Finally, it appears that there is a more efficient way of setting aside acceleration: namely, to declare that it may not exist! I quote two articles with the following titles: "An acceleration-free version of the clock paradox" (Low, 1990) and "Unaccelerated-returning —

Twin in Flat Space" (Brans and Stewart, 1973). These two articles present a very special version of the Clock paradox in choosing a close flat space for the traveler who comes back regularly to the stay at home. In fact, these models are very different from the original linear version of the problem. However, these models are used as proof that a Clock paradox may be created without acceleration. But some authors go too far and draw a general conclusion that acceleration is not a part of the problem. This is the definitive death of acceleration!

7.2. The Symmetry Breaking

One reads very frequently that the difference between the two observers is that the stay at home remains all the time in an inertial frame while the traveler is in an accelerated frame (or maybe two frames, one outbound and the other inbound). This is correct. However, if the acceleration time is negligible, this means that the traveler is almost never in an accelerated frame! He remains in inertial frames and consequently some scientists conclude that the two observers are in reciprocal positions in the two parts of the travel. We shall see later that some people ask: under these conditions, where is the symmetry breaking (although they do not use this formulation)?

From what we have said so far, there are two mechanisms which give us the symmetry breaking. However, in fact, they have the same origin, acceleration.

The first is the stopping and this implies the presence of acceleration. We have seen above that the observer who may stop is the one who remains at a smaller age upon the reunion of the twins. In fact, only the observer with the engine may cancel the relative velocity between the two observers and has the possibility of entering the frame of the stay at home. If one does not mention the stopping, one has only two observers in reciprocal positions. By reciprocal positions, we mean that all that was said about one observer taken at rest can also be said about the other also taken at rest.

The second mechanism that breaks the symmetry is the acceleration during the motion. The traveler is accelerated relative to the stay at home and consequently the rate of his clock is smaller.

But, as velocity is relative, acceleration is also relative. This means that each of the observers sees the other accelerated. However, there is a difference. The traveler is accelerated, because a force is applied to his rocket through the engine. For the traveler, the stay at home is accelerated because of the relativity of motion (see Edington (1923) for this difference in the accelerations). However, the traveler has a problem, which Einstein tried to solve: the traveler sees the stay at home accelerated without force! Einstein introduces a gravitational field, which at a first glance seems artificial. With the help of the principle of equivalence the reason for the introduction of this gravitational field is more convincing.

However, some scientists have felt the necessity to show that their model produces symmetry breaking and to introduce a specific mechanism. For example, length contraction was introduced but only from the point of view of the traveler who sees the earth–star distance shorter than that measured by the stay at home (Muller, 1972; Scott 1959). In fact, length contraction is a symmetrical effect and consequently it concerns two observers. Scott adds another reason for the asymmetry, namely that the star–earth distance is fixed relative to the stay at home and not the traveler. It is not a very convincing reason.

In the framework of the Special Theory of Relativity, the symmetry breaking is associated with the turnover, when the traveler changes his direction and comes back toward the stay at home. The simplest example is the Lorentz model. One finds some qualitative arguments invoking the role of the acceleration and the change of time in the traveler's clock in Stauton and van Dam (1980). However, one has to introduce explicitly the stopping at the end of the traveler's trip so that the two observers will be able to compare their ages.

In the framework of the General Theory of Relativity, when the principle of equivalence is used, the turnover is not necessarily associated with the solution. For example, the Born solution indicates only the acceleration at the turnover, but we have shown that it suffices that the traveler stops on the star and already his clock is behind the clock of the stay at home (if, on the star, there is a clock synchronous with the clock of the earth). In the Born model, symmetry breaking is associated with a gravitational field, which is equivalent to acceleration and is introduced only when the traveler is near the star.

When one considers Møller's solution and all solutions based on his work, one sees that the turnover does not play any role in the problem in spite of the solution of Born. This last solution may be easily adapted to the outbound trip only. As shown above, in all these models the acceleration is introduced explicitly and as a consequence, one has a real model of the Langevin experiment: the two observers are in the same frame at the end of the journey and are able to compare directly their ages.

7.3. The Exchange of Signals

This method consists of observers sending electromagnetic signals to one another. There are several articles on this method, which was very popular since acceleration was apparently not involved in the calculations. The observers send electromagnetic signals and count the total time for the signal arrival. The derivation developed below of the travel times through the signal exchange follows those of Tonnelat and Bohm.

We shall begin by taking the stay at home A who sends signals to the traveler B with frequency f_0. However, for B leaving the earth, because of the Doppler effect the signals that B receives have the frequency

$$f_a = f_0 \sqrt{\frac{1-\beta}{1+\beta}} < f_0. \tag{7.1}$$

On coming back to earth, the signals that B receives have the frequency f_r given by

$$f_r = f_0 \sqrt{\frac{1+\beta}{1-\beta}} > f_0. \tag{7.2}$$

We note that N is the total number of signals emitted by A, N_a is the number of signals B receives in the first part of the travel, and N_r is the number of signals B receives in the second part of the travel. One has

$$N = N_a + N_r, \tag{7.3a}$$

and

$$\frac{N_a}{f_a} = \frac{N_r}{f_r}. \tag{7.3b}$$

From (7.3a) and (7.3b), one can calculate N_a and N_r. One obtains

$$N_a = \frac{N}{1 + \frac{f_r}{f_a}} \quad \text{and} \quad N_r = \frac{N}{1 + \frac{f_a}{f_r}}. \tag{7.4}$$

The travel time of A is $\Delta t = \frac{N}{f_0}$ and the time of B is $\Delta t' = \frac{N_a}{f_a} + \frac{N_r}{f_r}$. The final result is

$$\Delta t' = \Delta t \sqrt{1 - \beta^2} < \Delta t. \tag{7.5a}$$

The travel time of the traveler B is shorter than the time of the stay at home A.

But now consider the case where it is B who sends signals to A: When B leaves A, the frequency of the signals that A receives is f_a given by (7.1), and when B comes back, the frequency of the signals that A receives is f_r given by (7.2).

The time of the trip of B is $\Delta t' = \frac{N}{f_0}$. During the separation of the observers (B leaves A), B sends $N/2$ signals and during the reunion (B moves toward A), B sends $N/2$ signals. One has for the stay at home A

$$\Delta t = \frac{N}{2f_a} + \frac{N}{2f_r} = \frac{N}{f_0} \frac{1}{\sqrt{1 - \beta^2}} > \Delta t'. \tag{7.5b}$$

Again the travel time of the traveler B is shorter than the time of the stay at home A. One sees that, irrespective to who sends the signals, the time of the observer B is shorter than the time of the observer A. Note that the two preceding calculations were made in the frame of the stay at home A.

Now, suppose that one considers this situation from the point of view of the traveler B as seen at rest. It is obvious that the results obtained will be identical, with the difference that it is for the stay at home (seen in motion by B) that the time of signal exchange will be shorter. These results, from the two points of view of the observers, are merely consequences of the reciprocity of the time dilation. One concludes that the signal exchange cannot give the age difference in particular, since there is no acceleration in the calculations.

7.4. The Three-Frame Model

The three-frame model is very attractive, since there is no acceleration in this model: the three mobile frames do not stop but move indefinitely. Consequently, the Special Theory of Relativity is the means to solve the problem.

The three-frame model has peculiar properties. In the end of this chapter, the solution of the model is presented, but for the time being the results are displayed in Table 7.1.

Consider first the stay at home. The time in his own frame is $(\Delta T)_1 = 2L/V$; it is his age. However, when he reads the time of the traveler, he obtains $(\Delta \tau)_1 = 2L/\gamma V$ which is shorter than his age. It results that $(\Delta T)_1/(\Delta \tau)_1 = \gamma$. In other words, the stay at home measures in his frame the famous ratio γ.

Now consider the traveler. In his own frame, the time is $(\Delta \tau)_2 = 2L/V$ and it is his age. The two observers have the same age as seen above. When the traveler reads the time of the stay at home, he obtains $(\Delta T)_2 = 2\gamma L/V$, which is larger than his age. The ratio

Table 7.1. Travel times as seen by the two observers from their own frames and the frame of the other.

	From the reference frame A	From the reference frames B/C
Travel time of the stay on the earth	$(\Delta T)_1 = 2L/V$	$(\Delta T)_2 = 2\gamma L/V$
Travel time of the traveler(s)	$(\Delta \tau)_1 = 2L/\gamma V$	$(\Delta \tau)_2 = 2L/V$

$(\Delta T)_2/(\Delta \tau)_2$ is again equal to γ. Both agree that their reading of the time ratios, each in his own frame, is equal to γ.

This sounds like the Langevin problem, but it is completely different since the ratio of the measured times is not the ratio of their proper times or their ages. One can understand why some authors made this confusion, since effectively the results of the three-frame model have something analogous to the original article of Langevin.

7.4.1. *Solution of the three-frame model*

We recall an important result of the Special Theory of Relativity we mentioned in Chapter 3. If a process occurs in an inertial frame S during a time τ, then in all other frames S_i moving with velocity V_i relative to S, the time of the process is given by a time τ_i that is different from τ.

To be more precise, we come back to the example of the use of the Lorentz equations we exposed in Chapter 3.

One considers a process composed of two events in the frame S'. One event is the beginning of the process and takes place in S' at $(x' = 0,\ t' = 0)$, and the other (which is the end of the process) takes place at $(x' = 0,\ t' = \tau_1)$, i.e., at the same place. The total time of this process in S' is τ_1. What is the time τ_2 of this process measured in S? From the application of the Lorentz equations, one obtains for the end of the process as measured in S, $x = \gamma V \tau_1$ and $\tau_2 = \gamma \tau_1$. In other words, the time measured in S is larger than that measured in S' since $\gamma > 1$, and it is obvious that for the observer in S the clock in S' goes slower than his own clock. But be cautious, the clock in S' does not go slower; it keeps a rate identical to that of the clock of the frame S. It is very often written that the clock in S' goes slower without mentioning that this is only as seen by S. Too often, the words "the clock goes behind" or "the clock goes slower" are taken literally and this is the source of very numerous confusions. We ask the reader not to forget this important remark to understand what we shall write now. We shall come back later to this point.

We shall consider the three observers and calculate their time readings during the whole process. The earth–star distance is L and the velocity of the travelers relative to the stay at home is V in absolute value.

We begin with the observer A, who is at rest on the earth. The calculation of his proper time is very simple. The traveler B travels a distance L with velocity V, and the traveler C travels a distance L with velocity V (in absolute value). The proper time of A measured in his frame is merely $2L/V = 2\tau$.

The frame B crosses the star and meets the frame C where the coordinates of B in the frame A are:

$$x_1 = L, \quad t_1 = L/V = \tau. \tag{7.6}$$

Now, we determine the reading made by the observer A of the time of the traveler B. It is given by the Lorentz equation. This corresponds for B as seen in the frame A:

$$\begin{aligned} x_2 &= \gamma(x_1 - Vt_1) = 0, \\ t_2 &= \gamma(t_1 - Vx_1/c^2) = V/(L\gamma) = \tau/\gamma. \end{aligned} \tag{7.7}$$

For the observer A, the time that the traveler B takes to reach the star is τ/γ. By reciprocity, for the observer A, the time that the traveler C takes to reach the earth is also τ/γ. In conclusion, for A, the time taken by the whole process for the frames $B + C$ is $2\tau/\gamma = 2L/\gamma V$.

One has to be very cautious about the meaning of the times of the observers B and C. They are not the times that they read by themselves at crossing the star; they are not their proper times but they are the times that are read by A on his own clock for the travel time of $B + C$.

Now, consider the frame C at rest. It sees the frame A traveling toward its position at the velocity V. The frame B disappears in the space.

The frame A moves toward the frame C until it crosses it, and when A begins its motion, the coordinates of A (measured by A) in

C are

$$x_2 = L, \quad t_2 = \tau. \tag{7.8}$$

Always when A begins his motion, the coordinates of A as seen from the frame C are

$$x_3 = \gamma(x_2 + V t_2), \quad t_3 = \gamma(t_2 + V x_2/c^2) = \gamma\tau(1 + \beta^2). \tag{7.9}$$

Now A crosses C and his coordinates in A are

$$x_3 = 0, \quad t_3 = 2\tau. \tag{7.10}$$

But his coordinates as determined by C are

$$x_4 = \gamma(x_3 + V t_3) + K = 0, \quad t_4 = \gamma(t_3 + V x_3/c^2) = 2\gamma\tau. \tag{7.11}$$

One has to add the constant K, as suggested by Einstein (1905), to obtain $x_4 = 0$.

When A meets C, his clock indicates his proper time 2τ, but for C the A clock indicates $2\gamma\tau = 2\gamma L/V$ that is different from his own proper time. This is something unusual and has some analogy to the Langevin traveler. One can understand why some people think that the three-frame model mimics sufficiently the original Langevin problem (Grünbaum).

One can verify that the constant K is equal to $2V\gamma\tau$ in (7.11). It is also possible to calculate the time needed for the frame A (from the point of view of the frames B and C) to come back to its original position and one obtains τ/γ. Thus, the time of the stay at home, from the point of view of B and C, is divided into three parts. Firstly, when A goes away, the time is τ/γ from (7.7), then the change of frame from B to C induces a change in the time $[\gamma\tau(1 + \beta^2) - \tau/\gamma]$, and finally, A comes back to C during the time τ/γ. One recovers the situation observed above, namely a jump in the time of A, always from the point of view of B and C. One can check again that the sum $2\tau/\gamma + [\gamma\tau(1 + \beta^2) - \tau/\gamma]$ is equal to $2\gamma\tau$, the total time as measured by the observers $B + C$.

As a final word about the three-frame model, we find a different way to write the discontinuity of the time of A as seen by $B + C$, $[\gamma\tau(1+\beta^2) - \tau/\gamma]$. It is possible to show by a simple calculation that

$$[\gamma\tau(1+\beta^2) - \tau/\gamma] = \frac{2VL}{c^2\sqrt{1-\frac{V^2}{c^2}}}. \tag{7.12}$$

This is the value of the time discontinuity K found in the Langevin model (see Chapter 4). Curiously, although the three-frame model is different from the original Langevin thought experiment, they have several things in common.

Summary of Part II

An important conclusion one can draw is that acceleration is an essential ingredient to understand the aging difference of the Clock paradox. Without acceleration, there is no departure, no change in the direction, and no landing on earth on coming back. The tendency to present the three-frame model as an equivalent of the Langevin experiment is purely and simply not correct.

Three solutions were presented in detail. The first (Born, Kopff, and Tolman) is characterized by a very short time for the application of the force and, consequently, the acceleration is present only for short times. In this solution, time dilatation plays an important role. However, the dissymmetry is explained as a consequence of the acceleration at the turnover. But this solution presents two disadvantages. First, it is good for only relatively small velocities of the traveler, and secondly, whatever the intensity of the applied force, the times of travel measured by the two observers are only a function of the velocity. This is a strange result that masks the role of acceleration.

The second solution is a solution without time dilation. The traveler activates his engine throughout the time of travel, once in one direction (toward the star) and once in the other direction in order to come back to the earth. In such a situation, the travel times as measured by the two observers is indeed a function of the strength of his engine.

The third solution is clearly the complete solution, which implies the presence of two factors: the time dilation and the strength of the engine, also called the acceleration.

Despite the opinion of several authors (Builder 1957) claiming that the Theory of General Relativity does not have a role in the solution nor has usefulness, it is essential in order to "understand" it. One puts quotation marks because the word "understand" has double meaning: through the equations, it is the quantitative meaning and through the physical laws, it is the qualitative meaning. A surprising aspect of the problem is that quantitatively, Special Relativity is enough to solve it but qualitatively one needs General Relativity.

The solution presented in Chapter 6 (the Møller solution) is entirely based on Special Relativity. One recalls the method. First, in the frame of the stay at home, one solves the problem of a mass under the influence of a constant force when the path to the star is divided into three parts. One obtains, through the dynamics equation of Special Relativity, the travel time of the traveler. It is a simple, elementary, and precise method.

Now, it is always possible in a reference frame S to know the proper time of another frame S' if, in S one knows the velocity of S'. In the present case, the velocity of the traveler relative to the earth is calculated; the second step is to determine the proper time of the traveler. The problem is solved and one has only to verify the aging difference.

But now why is the traveler younger? The answer is very simple: it is accelerated relative to the earth. Here the man who has control of the motion is the observer in the engine. His acceleration is a consequence of the force that the engine produces. In particular, it is only his frame which can enter the frame of the earth.

But effectively acceleration is reciprocal and the traveler sees the earth moving away with his acceleration. But he has a problem. He knows that there is no engine in the frame of the stay at home; therefore he concludes that this acceleration comes from a gravitational field.

To conclude, the Theory of General Relativity gives us the possibility of understanding qualitatively the Clock paradox.

PART III

DISCUSSIONS

Chapter 8

Physicists Out of the Mainstream

The Theory of Relativity has always been the subject of discussions and negations almost from the beginning (von Laue, 1912). Here I want to focus only on contestation from physicists. There were (and maybe always are) some groups of opponents who were not physicists but only scientists from other disciplines.

In his paper on the Clock paradox, Cochran (1960) mentions physicists who do not accept the Relativity Theory, and he makes this interesting remark: "there may be many [physicists who do not agree with Einstein] but only few have expressed their view in print". By this, Cochran suggests that the difficulty to accept the theory is more widespread than one can imagine.

Among the reasons invoked for discussing the Theory of Relativity, one finds the Clock paradox. Maybe it is a central problem in the Special Theory of Relativity, since it is almost always presented in the writings of critics. Einstein had good reasons to begin his imaginary dialogue (in his 1918 article) with a critic by questions about the Clock paradox.

The most famous debate about the validity of the Special Theory of Relativity was in the 50s and 60s of the preceding century. Dingle was at the center of this debate and in a long series of articles (in particular, in *Nature*), he presented his non-orthodox opinions. Because he was a physicist with a good record (he was Professor of Natural Philosophy at Imperial College, Head of the Department of Physics at Imperial College, President of the Royal Astronomy Society for two years, and Professor of History and Philosophy of Science at University College London), numerous physicists feel the need to give answers to his criticisms. It was a little bit curious that a physicist who wrote a book on Relativity in 1940 (*The Special Theory of Relativity*, 4th edition in 1961) became an opponent of the Special Relativity. Chang (1994) presents a long analysis of Dingle's attitude but does not discuss where his errors were.

A second debate was very short and Sachs was the physicist who wrote several articles on the Clock paradox denying the ageing difference. It was in the middle of the preceding century and I think that it was the last discussion of physicists about the Clock paradox. It is likely that the arguments presented by Dingle and Sachs will be found among other physicists but Dingle and Sachs were typical in their opinions.

One has to note the attractive nature of the debates and their vehemence after the publication of an article negating that the traveler is the younger one. Furthermore, in the Dingle and Sachs debates, no side was convinced by the arguments of the other side. However, one can learn from the errors of the critics and this is the purpose of this chapter. It is very easy to claim that the physicists who criticized the theory did not understand it.

Before Dingle became one of them, there were already opponents of the age difference. As an example one can consider Milne and Whitrow (1949), who wrote two articles to show that the two observers will have the same age when they meet again after the trip of one of them. The authors present two arguments. One is based on the concept of clocks. They ask: do "all natural processes which run naturally of themselves independently of what we may do may equally serve as clocks and give the same results"? It is well known that for Einstein and after him for the physicist community, the answer is "yes". However, for Milne and Whitrow, the answer is "no" and they make use of the concept of "physiological time", which is not equivalent to the "physical time". This new notion, in physics, of "physiological time" is presented without a definition, but for them, it is sufficient for excluding the age difference. The second argument is based on the symmetry between the two observers. Milne and Whitrow add that at the meeting, they will have the same age. Probably the reason for their opposition to the age difference is revealed by a combination of their two arguments. However, it is not very clear how they are connected. I am not aware of the reaction of other physicists to the opinion of Milne and Whitrow.

Dingle was associated with several debates. However, the one which is remembered is the Clock paradox and the validity of the Special Relativity. He participated in discussions about the cosmology model of Milne in the 1930s and 1940s (Gale, 2019). He was not the sole opponent of the model of Milne, and several other scientists were involved in this debate. An interesting aspect of this controversy is that it concerned not only physics but also philosophy of science. At the end of the 1930s, there was an exchange of articles in *Nature* with Campbell about the nature of time. In addition, in 1960 there was again an exchange of articles in the *British Journal of Philosophy of Science* with Grünbaum about the theoretical and philosophical basis of the Relativity Theory.

8.1. The Dingle Controversy

As mentioned above, Dingle wrote a large number of articles presenting his opinions and one can choose the article published in *Nature* in April 1956 (Dingle, 1956A) to begin with. The essential point of this article is to say that there is no age difference. Moreover, he explains:

> The two clocks in question, which would have continued to agree if they had not separated, show different times at reunion, something must have happened to one which did not happen to the other and nothing is in question except their relative motion. Hence they cannot show different times on reunion
>
> But those familiar with the customary phraseology of the theory will ask what is meant by the familiar statement that a moving clock runs slowly... It simply means that the reading is behind that of an imaginary stationary clock at the same place, which is synchronized with observer's stationary clocks... nothing at all happens to the moving clock...
>
> When the observers meet, again they are both at the same place and relatively at rest. Their judgment of simultaneity agree, and so their clocks agree.

It is always difficult to follow his train of thought, since Dingle introduced several subjects in his papers. Nevertheless, the argument of Dingle is very simple. The two parts of the journey are completely symmetric, since nothing happens to either of the observers. So, where is the asymmetry? He supposes that the turnover does not change anything. But we know from the works of several scientists like Born and Tolman that there is acceleration at the turnover and following the principle of equivalence, this induces changes in the travel time. Is it possible that Dingle was ignorant of the principle of equivalence? We shall discuss this point later but from his article of April 1956, it seems that he does not want to consider this principle.

Here it is worth considering the ironic remark of Dingle about an article of McCrea on the Clock paradox. First one has to come back to the McCrea article of 1951. To solve the problem, McCrea uses this length contraction:

> The stay at home R "plots" M [the traveler] as at distance $X = VT$ from himself. Owing to the Lorentz contraction, M at the same event plots R as at distance αX from himself, when $\alpha = (1 - V^2/c^2)^{0.5}$. During the interval t when R and M are both at rest in F [frame of R], they must each plot the other as at distance X. But as soon as M has again acquired uniform speed V, this time towards R, he again plots as at distance $\alpha X \ldots M$ moves out and back through distances X at speed V; total time $2T$. On the other hand, M's account of R's journey relative to himself is: R moves out through distance αX, then R 'cheats' by rushing out to distance X and back again to distance αX in negligible time and finally completes his return at speed V: total time $2\alpha T$. This is less than $2T$ and is the usual result.

The reaction of Dingle was "Cheating, indeed" and he adds:

> I know of no other example in the history of science in which fantastic propositions have been put as sober scientific truth.

Very frequently it is difficult to agree with Dingle, but in this case we do. To read that the distance RM is (for M) αX (at the end of the outbound path), suddenly changes to X, and comes back to αX when R moves again toward M is really cheating!

Just after his article, there were an article by McCrea (1956) and the answer by Dingle. McCrea proposes three answers to justify the age difference. One is the fact that the part of the travel at constant velocity can be broken into four different parts that he calls "free paths". By contrast, the stay at home has only one "free path". The second answer is that the traveler has an engine which gives him the possibility to move. The third is that the world lines of the two observers are different and hence the difference in the time.

One can wonder if these three answers could be satisfying for Dingle. His reply is very short. Firstly, the acceleration takes place in a very short time and so the subject of the engine is excluded. He adds that the remarks of McCrea about geodesics are "a needless generalization" and this is not related to the statement he made. It is clear that the introduction of the notion of "free path" is useful since it concerns only times when the engine is cut out, i.e., almost all the time. It is a pity that McCrea did not develop the role of the engine more fully. For Dingle, the Clock paradox is only a kinematical problem, but in fact, it is dynamic. A force is applied to the traveler, and it is necessary to consider it.

After reading the text of McCrea, one can only wonder why McCrea does not find a good answer to the real question: where is the asymmetry? Furthermore, one can again ask where the principle of equivalence is?

This exchange of articles provoked a number of articles, some approving Dingle, some disapproving him. The editor of *Nature* asked Dingle and McCrea to summarize their opinions about this debate and publish them. Their works were published on 29 September 1956, first that of Dingle (Dingle, 1956B) and, shortly afterward, that of McCrea.

The main point in the Dingle article concerns the acceleration. He replies only to people who disagree with him and he distinguishes two positions: (a) those who hold that the acceleration is irrelevant and (b) those who regard the acceleration as important. He put aside point (a), because without acceleration there is no Clock paradox.

About point (b), after a rather long analysis, he concluded that the time left for the acceleration is too short to have any influence and if one supposes that the acceleration may influence the rate of a uniform motion, all the "astronomical deductions are vitiated". The conclusion of his analysis is, as mentioned, that the acceleration is present but without effect. In brief, Dingle sticks to his position: at their reunion, the two clocks indicate the same time. The end of his article shows that some critics were suspicious about his mathematical skills. He closes by making some remarks about his knowledge in mathematics although he was an experimentalist.

About the answer of McCrea, the tone is not very sympathetic. We concentrate only on the part of his text concerning the acceleration and one cannot say that he is very fair. In order to be understood, he supposes that the turnover of the traveler journey is very fast. What does happen to the traveler? He would be killed. McCrea's intention is clearly to show that, if Dingle should be at the place of the traveler, he would be killed. This way of showing the influence of the acceleration is not very smart, especially because in his own model (1951 article), he gives a time t for the traveler to stop on the star! Secondly, he repeats his well-known argument that "the duration of the acceleration could be made arbitrarily short with other times involved". So what?

It does not seem necessary to make a summary of the McCrea article, since he does not give any answer to the basic Dingle question: where is the asymmetry? What is the mechanism of the symmetry breaking? Instead of giving an answer, he closes with a joke about the "cheating" of his stay at home:

> The curious details of M's plot of R's behavior when M treats himself as at rest are not really important, because M will discover that it is much simpler not to treat himself as at rest.

The impression that we get after reading these five articles is that we do not learn anything new. The real debate about the origin of asymmetry is masked by a lot of details that are really not relevant. So we are not astonished that the subsequent articles will present the same characteristics: each side speaks for himself in a dialogue of the deaf.

The same year in which he had a debate with McCrea, Dingle (1956C) published in the *Proceedings of The Physical Society* a long article with his solution to the Clock paradox. This time, he chose the model of the three frames, since there is no acceleration in this model. His solution is very complicated and he obtains a good solution; there is no age difference. This article inevitably received answers. We take Crawford's answer published in *Nature* (1957). In his answer, Crawford discusses Dingle's solution arguing that he made some errors. But the most interesting portion of Crawford's answer concerns the source of the asymmetry:

> In the clock paradox trip, the source of asymmetry is not the relative acceleration of the two twins; rather, it is the relative acceleration of each twin separately with respect to the 'third body' — the inertial (all frames in uniform motion with respect to the distant galaxies).

Of course, this answer did not correspond to the expectancies of Dingle and the same year (Dingle, 1957) published a more precise exposition of his views as an answer to Crawford.

Again, Dingle expressed his views in an article published in 1958 in *Science* as an answer to an article by McMillan (1957), also in *Science*. There is nothing new in Dingle's article: he repeats what he had published except for his mention of the 1918 article by Einstein. Now it is clear that he knows the principle of equivalence, but he declares without proof that it does not help to solve the contradiction. By contradiction, one can suppose that he hints at the solution of the Clock paradox.

After 1960, the situation changed. From this date, he made a new statement, namely that the Special Theory of Relativity is wrong. He published numerous articles with the hope that other scientists will agree with him but, on the contrary, he received only rebuttals. Upon the negation of the validity of the theory, very frequently he introduces in his articles the subject of the Clock paradox. This seems analogous to some obsession. He also made philosophical deductions about the role of the theory and the experiment (Dingle, 1964). The last paper was published in 1980, two years after his death, and presents his legacy (see Ronan, 1979).

In the present book, there is no place for a detailed analysis of the positions of Dingle nor why and how he evolved. It is certainly a very intriguing phenomenon and one can ask: why nobody succeeded in showing that he was wrong? Why, after the first article, nobody understood that the point was the symmetry breaking? It is evident that he had a clear idea of the concept of reciprocity between two inertial frames when it seems that so many scientists forgot it (think about the three-frame model analyzed in Chapter 6). I come back to this point in the following chapters.

This story leaves a bitter taste. Firstly, because most probably, Dingle suffered a lot and secondly, how is it possible that in physics, which is in principle objectivity, scientists do not succeed in reaching an agreement about a well-established theory?

8.2. The Sachs Controversy

In 1971, Sachs, who was Professor of Physics at the State University of New York, published in *Physics Today* (September 1971) an article in which he defends the view that at their reunion, the twins will have the same age. Contrary to Dingle, Sachs did not consider the Relativity (Special and General) as wrong. On the contrary, he thought that it is possible to show the equality of the ages by deduction from the theory itself.

His first argument is to claim that "[the] correlation between a fixed observer time measure on a moving clock and its aging process is not necessarily true". For him "the space–time parameters are a useful language to *express* the laws of nature, whereas the physical interactions *obey* the laws of nature". In other words, when an observer measures the time of a process in a frame moving relative to him, there is no one-to-one correspondence between the measuring time and the proper time of the process in the moving frame.

His reasoning goes as follows: (a) There is no complete solution of the Twin paradox even for those who accept the one-to-one correspondence mentioned in the preceding sentence. He criticizes the Tolman solution in that it is correct only for small velocities, (in which he is right) but he ignores the Møller solution. (b) He claims that, until now, the application of the principle of equivalence brought only a qualitative argument to support the age difference. This is not enough; a quantitative analysis is necessary. (c) I quote from his article: "It appears to me that two identical clocks, synchronized in a single proper frame of reference, should be synchronized in all future times of observation, even though they may not appear to be synchronized unless they are again in the same proper frame of reference". (d) A mathematical demonstration is presented in order to show that the variation of the proper time does not depend on the path in the space–time.

His concept of "proper time" implies that two clocks, once synchronized in one reference frame, will always be synchronized in that frame whatever the story of each clock. For him, it is analogous to the case of two identical rods that have their proper length in the same frame and when one is put in motion and comes back at rest in the original frame, they have again their proper length. In fact, as shown above, the situation is more complex and the encounter of two clocks is not a simple phenomenon.

He intended to give a proof that the proper time along a path in the space–time is path independent. For that, it is necessary to introduce the notion of distance in the space–time of the General Relativity. The four coordinates x, y, z and it with $i = \sqrt{-1}$ are written as four equivalent coordinates x_1, x_2, x_3, and x_4. The squared differential of the proper time is in general

$$ds^2 = \sum g^{\alpha\beta} dx_\alpha dx_\beta.$$

However, Sachs preferred the differential of ds without taking the square root of ds^2. For this, he wrote the differential ds as a complicated expression involving quaternions. From this point onward, it is difficult for one to follow his reasoning without being a mathematician expert in the General Theory of Relativity. But we can mention Rodrigues and Rosa (1989) who have presented a discussion of the argument of Sachs to show that he was wrong.

The most interesting aspect of the debate is not only the views of Sachs but also the reactions of physicists of the mainstream. These were published in *Physics Today* (January 1972) and show that the dialogue was again a dialogue of the deaf.

A group of articles recalled time dilation, like the lifetime of the muons, as an explanation for the Clock paradox. Obviously, there is no place for such a recall. Others made some errors, such as to think that the acceleration is absolute. There were also articles ignoring time reciprocity. Finally, some scientists discuss details of the Sachs demonstration without exposing clearly where the error is. However, among the critics of Sachs, there is an astonishing article by Greenberger titled "The reality of the twin paradox effect". Did he mean that the paradox is true and that there is no solution? Surely not, but one has here, a curious slip.

In 1973, Sachs published a criticism of the article by Wu and Lee (1972) who present a solution entirely based on the General Relativity. The principal argument of Sachs was that the solution of Wu and Lee is only an approximation as an application of the scheme of the article of Einstein of 1918.

8.3. Dingle Versus Sachs

A strange conclusion that one can draw from this debate is that supporters of the age difference did not succeed in presenting answers that could convince Dingle and Sachs. But the most curious aspect of these debates is that, in fact, many answers were really not convincing. One has the feeling that many scientists reasoned as follows. Step 1: Time dilation is the source of the age difference, as shown by the Lorentz transformation. Step 2: Acceleration is irrelevant. Step 3: It is well known that the traveler is younger at the end of the journey. Step 4: The ratio of the travel times is γ. *De facto*, the Clock paradox is a taboo; it is forbidden to cast doubts on the age difference solution. Sachs (1974) himself wondered how so many physicists were angry with the positions of Dingle and by his own positions.

In his article of 1974, Sachs developed his ideas again, and particular attention must be paid to his interpretation of the Lorentz transformation.

Thus, according to the logical structure of the theory of relativity, the inequality of time scales in relatively moving frames of reference does not refer to any asymmetry in the evolution of physical processes in the respective coordinate frames such as the unwinding of springs of clocks that are in relative motion...

In the same sense, the Lorentz transformations are not physical cause–effect relations. Hence, in the example mentioned above, the relation of the contracted time scale T_A to the time scale T_B is not a cause–effect relation that implies a physical effect...

It follows from this analysis that so long as there is no extra force entering the system, to physically act on the moving matter, but not to act on the matter in the observer's frame of reference (or vice versa) then there can be no asymmetric aging — according to the logical structure of the theory of relativity itself!

The essential point of Sachs is that the Lorentz transformation is not a cause–effect relationship. In other words, one cannot conclude about the presence of a physical effect in a given system by only using the Lorentz transformation. Therefore, in all those models of the Clock paradox that avoid acceleration, there is no asymmetric aging.

However, an astonishing point is that Sachs considers only the cases without force, i.e., without acceleration. He had already mentioned the book by Tolman but, like Dingle, Sachs seems to consider the principle of equivalence irrelevant. His target public remains those who attribute the Clock paradox to time dilation. Only in his article of 1973 does Sachs mention the principle of equivalence, but his conclusion is unchanged: the two observers have the same age. It is remarkable that among the answers to Dingle and to Sachs, there is not a single answer in which the role of acceleration is analyzed.

Maybe one can see Dingle as a victim. At the beginning, he asked a good question about asymmetry. However, he did not receive any correct answers. Finally, he thought that the theory is wrong. Interesting enough is the opinion of Sachs (1974) about Dingle. Although he did not share Dingle's view on the Theory of Relativity, Sachs expressed his sympathy for his independent mind.

Chapter 9

Questions With and Without Answers

There are several questions that are impossible to ignore and must be asked. The simple fact that I did not find them expressed in any philosophical articles is in itself very astonishing.

The first question is: why did some scientists forget reciprocity? Reciprocity is the basis of the Clock paradox but many articles give the solution by applying the time dilation only to the stay at home. Sometimes, it is justified by invoking two arguments. First, it is said that the traveler is accelerated and this makes the difference. But curiously, no clear conclusion can be drawn from this argument when one finds that immediately after presenting this claim, it is written that the acceleration time is so short that it does not play any role! Secondly, some physicists mention that after the turnover, the traveler is now in a new frame and this is the source of the difference.

The second question concerns the disappearance of acceleration. For what reason is the acceleration eliminated? As mentioned several times, it is an almost universal hypothesis to imagine that the time of application of the acceleration is very short. Why? This position is far from being innocent. We saw in preceding chapters that the role of acceleration is essential in order to understand the problem of the Clock paradox.

The third question is: why is the Møller solution known only by a minority of scientists? The book by Møller is frequently quoted but not his solution.

The fourth question is: why are there so many solutions?

I did not find answers to these questions, since apparently they were never asked. It is possible that, in view of the huge literature on the subject, I missed some important articles. I shall try to imagine answers and I am not sure that they are really good ones. But the goal is to attract the attention of the reader on this aspect of the phenomenon. It is obvious that these questions are related and consequently one cannot answer them one after another.

We begin with an important remark physicists do not dare to express. The story of the Clock paradox is unbelievable; it contradicts violently our concept of time. How is it possible that the traveler could be younger? In addition, what may be more difficult to swallow is the result that for the traveler, the time of the travel is shorter than the time that light takes to reach the star in the stay at home frame! Psychologically, there is a real problem that nobody wants to raise. For Einstein, the aging difference is a curious consequence of the theory, but it is not known whether all physicists share this attitude. One can say that it is a real dilemma and there is no solution: in our daily life, we have a simple feeling of the time, and the theory (that one cannot dispute) presents absurd results. It is possible that there is some kind of unconscious repression. I suggest the reader not to forget this point in the following text.

9.1. The Time Dilation and the Time Reciprocity

I do not know who introduced this expression, but it is not present in the first article of Einstein of 1905. As is well known, time dilation appears when one uses the Lorentz transformation. In the frame S, there is an event in (x, t), and this event is also observed in a frame S' in uniform motion relative to S with velocity V. What are the results of the measurements of this event made by the observer in S'. The answer is given by the Lorentz transformation:

$$x' = \frac{x - Vt}{\sqrt{1 - \frac{V^2}{c^2}}}, \quad t' = \frac{t - \frac{Vx}{c^2}}{\sqrt{1 - \frac{V^2}{c^2}}}. \tag{9.1}$$

Now if $x = 0$, the time t' measured is longer than t, $t' = \gamma t$. For observers in S', the event took a longer time than that for the observers in S. In other words, the time in S is smaller than the time measured in S'. Hence, the expression "time dilation". I quote two authors of textbooks:

McCrea (1951):

> Hence the time intervals registered by a clock appear to an observer with respect to whom it has a uniform velocity V to be lengthened by a factor $\sqrt{(1 - V^2/c^2)}$ or in brief *a moving clock appears to go slow*. This generally alluded to as the phenomenon of *time dilation.*

Born (1962):

> The same remarks apply to the relativity of time. An ideal clock has always one and the same rate of beating in the system de reference in which it is at rest. It indicates the "proper time" of the system of reference. Regarded from another system, however, it goes more slowly. In such a system a definite interval of the proper time seems larger.

These two quotations are enough, because one can find similar definitions in all textbooks. The novelty of Einstein is in saying that this effect is not apparent. In each frame, the time that observers measure is their time. They do not have another time. We call the effect of measuring different times, depending on where the time is measured, a parallax effect. We have to insist that the word "parallax" is not equivalent to the word "apparent". The measurement of a physical quantity is something objective, even if this measurement gives different results when measured in different frames. However, the time t of the event in S is its proper time, since it is determined by the clock at rest in this frame.

In the literature and textbooks, there are numerous discussions of the "reality" of the measurements of time in different frames. The classical proof of time dilation is the lifetime of the muons. The lifetime of this particle is much larger for observers on the earth when its velocity is close to the light velocity. The measurements of this lifetime (in fact, the half-life) are the proof that the measurements are objective and, consequently, real.

Unfortunately, an ambiguity is sometimes found in some textbooks or articles. The conclusion of time dilation is that the velocity slows down the clocks and one can quote Einstein himself:

> As judged from K, the clock is moving with the velocity V, as judged from the reference body, the time which elapses between two strokes of the clock is not one second but $\frac{1}{\sqrt{1-\frac{V^2}{c^2}}}$ seconds, i.e., a some larger time. As a consequence of its motion the clock goes more slowly than when at rest.

When one reads this full passage, it is obvious that the clock goes more slowly as observed from a frame moving with velocity V. But very frequently, only the last sentence is quoted, out of its context.

One has the feeling that sometimes Einstein did not always use a very precise language, and owing to his eminent position, people had a tendency to accept his writings without checking. One can imagine the great weight of a sentence such as "Einstein wrote..." or "Einstein said..."

One can find numerous examples of such absence of precise formulation. People want to make it short, and as a consequence, their writings are very ambiguous. As an example (among a large number of such examples), Arzeliès (1966), after explaining correctly what is time dilation, writes:

> This phenomenon is often referred to as *time dilation*; we also say that *a moving clock goes slow.*

This kind of formulation has catastrophic consequences, because many scientists merely adopt the simple conclusion: a moving clock goes slower. It is very common to read that the clock of the frame S', which is moving relative to a frame S, has a different rate than the clock of the frame S. However, unfortunately this is not correct.

For example, in the article of Frye and Brigham (1957), one reads

> ...by relativist principles, at that speed [99.999995% of the light velocity] his watch, [of the man traveling at this velocity] his heart beats, in fact his whole metabolism slows down to a rate determined by the formula $t_A = t_B \sqrt{1 - \frac{V^2}{c^2}}$

The authors state very clearly that the velocity itself makes the clocks slower. This error is very frequent and we shall see below its devastating effects. It is obvious that the concept of velocity without specifying relative to what it is defined is nonsense in the Einstein Relativity and, secondly, all that Frye and Brigham wrote is not correct, because a constant velocity has no effect on the metabolism. It is the principle of relativity itself which is forgotten.

To make it definitely clear, one has to mention that the clock of S and the clock of S' have the same rate, each in its own frame. This means that, for example, the same chemical reaction which takes place in both frames is measured by the two clocks (in S and in S') as having the same time τ. This is the exact meaning of two clocks having the same rate. Now, observers in S who measured the time of the reaction in S' finds that their clock indicates a time larger than τ, more exactly $\gamma\tau$. But one cannot conclude that the S' clock has a different rate. The measurements of the same chemical reaction (that is taking place in S') made by observers in the two frames S and S' are different.

Hence, it follows that it is incorrect to say that the motion of the clock relative to some reference frame destroys the synchronism between the moving clock and the reference frame. It is the departure and stopping of the moving clock that induce a change in the synchronism, because these two steps imply acceleration.

The conclusion that a process has no definitive time duration (since it depends on that frame in which it is determined) may be disturbing, because our mind tends to attribute to a process its specific duration. The Theory of Relativity asks us to modify our intuition. If one accepts this modification asked by the Theory, one will be ready to accept the apparently absurd results of the Clock paradox.

The reciprocity of the time measurements is well known. If there is an event (x', t') in S' the measurements (x, t) of the event in S are given by the other form of the Lorentz transformation:

$$x = \frac{x' + Vt'}{\sqrt{1 - \frac{V^2}{c^2}}}, \quad t = \frac{t' + \frac{Vx'}{c^2}}{\sqrt{1 - \frac{V^2}{c^2}}}. \tag{9.2}$$

From equations (9.1) and (9.2), one finds (what we mentioned just above) that the same temporal process which occurs in both frames S and S' is viewed reciprocally by the observers. Those in frame S see the process in their frame taking a time τ and the process in S' taking a time $\gamma\tau$. Reciprocally, those in frame S' see the process in their frame taking a time τ and the process in S taking a time $\gamma\tau$.

Very frequently, it is said that "the clock in S' is behind the clock in S" and that "the clock in S' is behind the S clock". How does one understand these two expressions?

The answer is very simple: one has to suppress the word "clock" and speak only about the process. Even if one considers an event characterized by its coordinate x and its time t, it may be seen as the end of a process that began at the point $x = 0$ and $t = 0$. As mentioned several times (and also by Langevin himself), a process that occurs in the frame S is measured with a longer time in all other frames S' that travel relative to S. But the sentences "The clock in S is behind the S' clock" and "The clock in S' is behind the S clock" cannot be taken literally. It is better not to use these expressions.

9.2. Forgetting Reciprocity

Despite being the basis for the formulation of the paradox, reciprocity is often forgotten. In many textbooks on the Theory of Relativity (Special and General), the Clock paradox has already been introduced in the chapters on Special Relativity. The authors give the well-known solution to the problem with, as is habitual, a very short time of acceleration: for the stay at home, the travel time is L/V and that for the traveler $L/\gamma V$ with no other explanation than time dilation. In the part of the text devoted to General Relativity, the principle of equivalence is introduced and one calculates the point of view of the traveler.

In the book by Møller about the Clock paradox, one can read:

> Consider two standard clocks C_1 and C_2 placed together at the origin of an arbitrary inertial frame of reference. At the time $t = 0$ the clock C_2 is set in uniform motion along the x-axis with velocity v. At the time $t = t_P$ it has reached a point P on the x-axis

and according to (2.6) [Lorentz transformation], it will record the time $t_P(1 - v^2/c^2)^{0.5}$. Immediately after arriving at P, C_2 is sent back to O with velocity $-v$. It arrives when the clock C_1 records the time $t_1 = 2t_P \ldots$ The clocks C_1 and C_2 record times t_1 and t_2 respectively connected by the equation

$$t_2 = t_1(1 - v^2/c^2)^{0.5}.$$

But where is the paradox? The author does not consider the case where the clock C_2 is seen at rest and C_1 has moved. In such a case, it is the clock C_1 that has passed a shorter time than the clock C_2.

It is very astonishing that in numerous articles using the solution on time dilation alone, no mention is made about a mechanism of reciprocity breaking. It is obvious that the sole mention of time dilation does not solve anything if one does not mention some other mechanism. We have mentioned above that the departure and the stopping of the traveler need to be introduced into the solution. In other words, acceleration is compulsory for the solution.

I do not have any satisfactory explanation for the "oblivion" of the reciprocity and I shall relate it to some other aspects of the phenomenon.

In the scientific community and among the amateurs of the Relativity, there are two well-established results concerning the Clock paradox. Firstly, the time of application of acceleration is reduced to an infinitesimal time and, secondly, the ratio of the two travel times is $\sqrt{1 - \frac{V^2}{c^2}}$. So, for authors willing to work on the problem, there is a completely new and strange situation. They want to propose a solution to a problem when the final answer is already known!

So how does it work? First, there is a decision one has to make: Special or General Relativity? Since apparently the problem does not involve acceleration (it was already excluded), the choice of Special Relativity seems logical. Consequently, it is natural to adopt the time dilation as the physical origin of the age difference. Now, one can have a wide range of possibilities, and even repetition of already published articles and solutions. However, the problem of reciprocity breaking remains. Thus, there are three ways to answer this question. One way is to ignore reciprocity in calculating the time dilation only for the stay at home, the other is to use a false argument (like length contraction), and the third is not to consider proper times like in the three-frame model.

Maybe one can go a little further and ask if the approach based solely on time dilation is not the consequence of application of the Lorentz Relativity. It seems that for some scientists there is some confusion between the Lorentz Relativity and the Einstein Relativity. As is well known, in the Lorentz theory there is a preferred frame, the one of the ether. Thus, the stay at home is in this preferred frame and, consequently, there is no reciprocity. In this theory, one can say merely "A moving clock goes slower". Since no mention is made of different frames, one may understand that the clock goes slower relative to the preferred clocks associated with the ether. Thus, there is no need for reciprocity and the effect is a true one.

One may wonder why the Lorentz Relativity Theory is still adopted by some scientists. The most well-known among them is John Bell with his paradox (2001). This seems strange, because frequently the Lorentz theory is viewed only as a step toward the Einstein Relativity. Lorentz found his famous transformation, but the majority of scientists prefer the Einstein interpretation.

In a very interesting article, Bell proposes to teach Relativity through the Lorentz version:

> We do not need to accept the Lorentz's philosophy to accept a Lorentzian pedagogy... And it is often simpler to work in a single frame rather than to hurry after each moving object in turn...
>
> But in my opinion there is something to be said for taking students along the road made by Fitzgerald, Larmor, Lorentz and Poincare. The longer road sometimes gives more familiarity with the country.

Bell wrote the above passage about a paradox called the Bell paradox. It could be interesting to examine in more detail the paradox of Bell in order to understand his point of view.

One considers two spaceships B and C moving in the direction $x > 0$ with a constant distance between them. They are accelerated by the same acceleration. Consequently, they keep the same velocity (which increases with time) and the same distance between them. A wire is placed between them from the nose of B to the tail of C. All these data concern the frame A. The prediction of the Theory of Relativity is that, during their travel, the wire will be broken. How is it possible?

Following Bell, the reason is because in the Lorentz version of relativity, there is length contraction between objects but not between isolated points in space. One recalls that for Lorentz the length contraction is a real effect, analogous to the length change of the rod when its temperature is changed. Consequently, in the frame of the spaceships, due to the Lorentz transformation, the distance between the spaceships will be constant, but length contraction exists for the wire. It experiences a stress, which increases with time until this stress will become large enough to break the wire. This explanation is simpler and one can understand why Bell prefers the Lorentz Relativity, since an explanation with the help of the Einstein version of Relativity is more complex.

Effectively one obtains the same result by a completely different way in the Einstein Relativity. It is possible to consider the position of the spaceship B in the frame of the other spaceship C (Semay, 2006) and to calculate the distance between them. The result of the calculations is that this distance as seen by C increases with time as a function of their common acceleration in the rest frame A. Consequently, there is a stress in the wire and, finally, it will be broken.

In the two solutions, use was made of the Lorentz transformation and it remains that observers in the rest frame A see that the wire is broken suddenly without explanation. Here we encounter Sachs for whom the Lorentz transformation cannot play the role of a cause–effect phenomenon. It is likely that Sachs would believe that the wire would not be broken.

A lesson that one can draw from the debate on this paradox (and here the literature is huge) is that it is very easy and maybe preferable to pass from the Einstein version to the Lorentz version of Relativity. In the case of the Bell paradox, we are concerned by the length contraction and in the case of the Clock paradox by the time dilation. Maybe one has an explanation for the forgetting of the time reciprocity, because as Bell explained, some ways are easier to grasp than others. The Bell solution can be explained qualitatively without equations and is easy to remember when the second solution cannot be explained without the machinery of the Relativity and one cannot propose a qualitative interpretation. One has something analogous to the Clock paradox. It is very easy to remember time dilatation without the necessity to introduce acceleration and the principle of equivalence. The possibility to forget time reciprocity is very high and that is what happened.

Time reciprocity is not a simple concept and one can see the sad effect in the case of Dingle. When he began to think that the Theory of Relativity is wrong, he based his view on an "unstoppable" argument, which can be stated as follows. When considering two frames, S and S', with relative velocity V, one says that the clock of S' is behind the S clock and at the same time, one says that the clock of S is behind the S' clock. Such a situation is absurd, since how can two clocks may be together and at the same time one behind the other? The conclusion of Dingle is that the theory is wrong.

9.3. The Acceleration

Who read the original article of Langevin? It is obvious that very few scientists read the article of Langevin and one can understand why. The journal in which it was published was not very easy to find and secondly it was written in French. The authors who published articles on this subject felt that they behaved honestly in quoting the original source from the journal *Scientia* without having read it. It remains that there is a lack of professionalism in such a situation. In fact, the large majority of scientists knew the paradox only by "rumors".

For the authors quoted above (Frye and Brigham), the idea of the thought experiment of the twins is due to ... Gamow! Today, there is no excuse for ignoring the original article of Langevin. The translation in English has been published by *Wikisource* since 2010 and may be read directly online.

Why is it so important to read the original Langevin article? One recalls that Langevin does not mention at all that time dilation is the cause of the aging difference. He insists (and with humor) that the origin lies in acceleration. Thus, it would be possible to find many more scientists who are ready to study the problem in this manner. Maybe time dilation should remain at its own place and should be seen as having a second role only.

The fact that acceleration was excluded is not a physical question but a psychological one. It is not necessary to read the huge literature, but only read attentively the book by Marder (1971) devoted to the Clock paradox. At the end of the book, there is a section titled "Paradox revisited", where there is a discussion as to whether the introduction of acceleration adds something to the solution of the problem. First, there is confusion between the introduction of acceleration and the use of the General Relativity to solve the problem. The author presents a strong criticism of the scheme of Einstein in his 1918 paper, declaring that it is not useful, since the Special Theory of Relativity gives the solution. This solution is clearly based solely on time dilation. Effectively the conclusion of the book is that time dilation is the solution:

> We can see that the general theory adds little to the interpretation of the Clock paradox in the absence of real gravitational fields... Fortunately, the amount of experimental confirmation of TIME DILATION is increasing rapidly. As the phenomenon at last becomes a familiar feature in scientific observations, the 'paradox' of asymmetrical ageing will cease to intrigue or perplex.
> (The capital letters are my own transcription)

So, as is clear from these last lines of the book, time dilation is the source of the asymmetry.

But in Chapter 3 of his book, Marder analyzes the case of a constant force during the whole trip, positive in the outbound path and negative in the inbound path, i.e., without time dilation. And he finds, as expected, that the traveler's time is shorter than the time of the stay at home. Consequently, it was expected that in the end of the book, the author will conclude that the origin of the asymmetry comes from acceleration. It does not come to the author's mind to associate the two parameters: time dilation and acceleration. Why?

The answer cannot be anything else than a psychological reason. It is possible to propose a tentative explanation. The first point is that the General Theory of Relativity is not very well known to many physicists, while people believe that they understand the Special Theory of Relativity. The General Relativity being not well known, there is some feeling that in order to solve the Clock paradox problem the Special Theory is a better way. But there is no serious reason for that. In addition, it is particularly gratifying to discover that the two approaches give the good answers but some people do not care. The second point is the notion of inertial frames which move one relative to the other at constant velocity. In the mind of some scientists, this means that there is no place for acceleration in the Special Theory. The best way is to exclude it by taking a very small time for acceleration.

One may apply this argument to Marder, who forgets at the end of his book the role of acceleration. Something like a Freudian repression.

The success of the three-frame model is clearly related to the absence of acceleration in this model. Are some people afraid of introducing acceleration? It seems that the role of the acceleration in the Special Theory of Relativity is not very well understood.

9.4.   The Multiplicity of Solutions

My interest in the Clock paradox comes from the huge number of publications on this subject. It is impossible to give the precise number of articles discussing the problem and proposing solutions. Maybe several thousands? A non-negligible number of these articles are merely repetitions of already published articles. It is unlikely that they are cases of plagiarism. It is more likely that, in view of the number of journals in which these articles are published, the authors (and the referees!) do not know that these ideas have already been published. The journals where the articles were published belong to different disciplines and this may explain (only partly) the multiplicity of articles: physics, history of sciences, philosophy of science, engineering, and also popular scientific works. Furthermore, the Internet is a common platform to publish more and more articles. It seems that the Clock paradox has a puissant power of attraction. It is likely that the first article was that of Lorentz in 1914 and to date (2019) several works have appeared each year.

The following are the basis for the solutions but only some of the references are given: the Lorentz transformation (Rossen, 1964; Müller, 1972); the Minkowski diagram (Rossen, 1964) or the space–time diagram (Chang, 1993); the Doppler effect and the exchange of light signals (Bohm, 1965; French, 1968); the three brothers or the three clocks (Grünbaum, 1954).

Shuler (2014) has provided the most extensive list of solutions that I know of. He made an inventory of articles published in the *American Journal of Physics* from 1957 to 2008. He found 21 different solutions!

At the same time, one has to add that many of the solutions are very complicated. For example, if in the original formulation, there are only two twins, in some articles, one finds three, four, or more brothers! Some articles are difficult to read because they are too complicated. Some articles present several solutions in the hope that the things will be clearer.

The first remark is the difficulty to accept the apparently absurd result. Physicists are used to the theory, but a student who hears for the first time this conclusion given by the lecturer, has a real difficulty. How is it possible? He does not have any intellectual tool to understand it. In fact, it is very likely that he "accepts" the result (because the lecturer has enough authority) but without conviction. The student remains in a "schizophrenic" attitude: on the one side, he cannot dispute the theory and must accept it, and on other side (or more precisely inside himself), the result seems beyond reality. Psychologically, he cannot internalize it. With time, the student becomes used to this kind of physical result (like the wave–particle duality) and apparently there is no more problem.

It is likely that this is so for some cases like the wave–particle duality, since nobody ever met an electron as a particle and at the same time as a wave. This apparent contradiction on wave or particle does not bother us in our daily life. This contra-intuitive concept does not interfere with our daily life. But when time is involved, it is a different story.

Our hypothesis is that the dilemma is not completely removed but remains hidden in the subconscious. And this may be the beginning of an explanation for the multiplicity of solutions.

For some scientists, the existence of this hidden dilemma may induce them to look for their own original solutions to the Clock paradox. In this way, they can convince themselves that they have understood it, since they found a new approach. Some solutions are really complicated (Field 2008) but this does not deter some scientists from going on.

In other cases, some scientists propose in the same article several solutions to the Clock paradox. The need to show that there are several ways to explain the aging difference may be a means to get reassured that the theory is correct in spite of its apparent absurdity.

We did not find another method than the appeal to the subconscious for grasping the phenomenon. We think that it is something strange that so many articles begin with the same two remarks: firstly, it is indicated that the Clock paradox is an old and not very well understood problem, and secondly, the author claims that he has a new idea to make things clearer. This kind of naivety may be explained as the existence of some hidden conflict the author tries to reconcile within himself.

9.5. Concluding Remarks

Why is the Møller solution (we recall that it is based solely on the Special Theory of Relativity) not well known except by a minority of scientists? We have no answer to this question and we leave it unanswered.

Finally, we only wish to note that we have directed our enquiry to the side of the unconscious mind. Is this the correct way? It appears as a plausible explanation, since, at first glance, no rational explanations is evident.

Chapter 10

Bergson and the Relativity

On peut être un physicien éminent et ne pas être
exercé au maniement des idées philosophiques.
La philosophie, comme le reste, a besoin de
s'apprendre.

One can be an eminent physicist and still not be
experienced in dealing with philosophical concepts.
Philosophy, like all subjects, has to be studied.

Henri Bergson

10.1. The Philosopher Bergson

Henri-Louis Bergson (1859–1941) was a French of Jewish origin philosopher who was influential in the tradition of continental philosophy, especially during the first half of the 20th century until World War II. Bergson is known for his argument that processes of immediate experience and intuition are more significant than abstract rationalism and science for understanding reality.

He was awarded the 1927 Nobel Prize in Literature "in recognition of his rich and vitalizing ideas and the brilliant skill with which they have been presented". In 1930, France awarded him its highest honor, the *Grand-Croix de la Legion d'honneur*.

Bergson's immense popularity created a controversy in France, where his views were regarded as opposing the secular and scientific attitude.

The above short description of the French philosopher Bergson is extracted from Wikipedia, which devotes a long article to him. It may seem strange that there is a place in this book for a philosopher. Obviously, Bergson and Einstein are, each in their domains, exceptional personalities, but apparently, the Bergson philosophy does not have any connection with physics. In fact, the only link was the concept of time.

In philosophy, the concept of time is one of the most investigated concepts. In the academic year 1902–1903, Bergson taught a course on "History of the idea of time" at the *Collège de France.* The content of this course was published after his death: Bergson revisited the philosophies of Plato, Aristotle, Descartes, Leibnitz, and Kant.

He was one of the people who gained a knowledge of the Special Theory of Relativity from the lecture and the article of Langevin of 1911. This new conception of time and space was for Bergson something that merited deep thought and so indeed at the meeting of Einstein with physicists and philosophers in April 1922, Bergson gave his first presentation about the difference between the concept of time for a philosopher and that for a physicist. As mentioned in Chapter 2, the answer Einstein gave did not show any sympathy for philosophers. Nevertheless, Bergson was not discouraged; in the autumn of the same year, he published "Duration and Simultaneity" with the subtitle "With reference to Einstein's theory". Immediately after its publication, debates began and people adopted different positions. The most important group was that of physicists, who discovered the "errors" made by Bergson and tried to convince him that he was wrong. However, unexpectedly, Bergson did not accept their criticisms and went on with his "errors". Philosophers were embarrassed, because they did not have the tools to judge the quality of the book.

The book was reprinted several times, but after 1931 it disappeared from view. It seems that the book was completely forgotten, and it was only recently that a new interest in the book emerged, especially among the contemporary French philosophers (During, and Deleuze) and the French physicist Levy-Leblond.

An English translation was published in 1965 with a preface by Dingle. We shall see below why Dingle wrote this introduction.

10.2. The Content of "Duration and Simultaneity"

It is worth to begin reading the book with the preface in which Bergson reveals his intentions. The first words are:

> A few words about the origin of this work which will enable the reader to understand its purpose. We began it solely for our own benefit. We wanted to find out to what extent our concept of duration is compatible with Einstein's views on time. The admiration for this physicist, our conviction that he was giving us not only a new physics but also certain new ways of thinking, our belief that science and philosophy are unlike disciplines but are meant to implement each other...

We can understand upon reading the preface that the prime intention of Bergson was to write for himself, but he later thought that it could be of interest for other philosophers to gain knowledge of his thinking. Although some critics (Sokal and Bricmont, 1998) consider it a book on physics, this is surely not the intention of Bergson, who wanted to write a book on philosophy. It is only from this point of view that one can understand Bergson.

The first two chapters are a summary of the Special Theory of Relativity: in Chapter 1 titled "Half-relativity" the Lorentz version is introduced and in Chapter 2 titled "Complete relativity" the Einstein version is introduced. The fundamental point for Bergson is the reciprocity in time and distance between two inertial frames moving with a constant relative velocity. In addition, he insists on the absence of absolute motion, since there is no particular reference frame for all the other frames. The first two chapters are presented for those who do not know the theory, and the content is taken from some textbook that Bergson does not indicate. However, the goal is also to emphasize the reciprocity between two inertial frames moving relative one to another with velocity V.

The next two chapters of the book are essential for his approach. Chapter 3 titled "Concerning the nature of time" describes in detail the conception of time of Bergson. We quote below sentences that contain the very essence of his idea of time:

> To tell the truth, it is impossible to distinguish between the duration, however short it may be, that separates two instants and a memory that connects them, because duration is essentially a continuation of what no longer exists into what does exist. This is real time perceived and lived. This is also any conceiving time, because we cannot conceive a time without imagining it as perceived and lived. Duration implies consciousness; and we place consciousness at the heart of things for the very reason that we credit them with a time that endures.

The most important idea is that real time is time which is perceived and lived. One can say that it is the definition of a "real time". One has to remember this definition each time Bergson mentions the concept of real time. However, what is the time that is not real? The answer is in Chapter 4, "Concerning the plurality of times".

Chapter 4 intends to analyze the times measured by the observer in the frame S for events that take place in the frame S'. This is the plurality of times. The S observer (who is a physicist) determines different times from different frames S' at different relative velocities and this plurality needs an interpretation. For Bergson, these times are not real but only virtual. These virtual times are "attributed" by the physicist in S to the events in S'. However, these times are not lived; the S observer is not conscious of them. In fact, these times do not correspond to the definition of real time.

> Thus, to sum up, whereas the time attributed by Peter [the S observer] to his own system is a time he has lived, the time he attributes to Paul's [the observer in the moving frame S'] is neither a time that either Peter or Paul has lived, nor a time that Peter conceives as lived or as capable of being lived by a living conscious Paul.

To the careful reader, things are clear and Bergson avoids all ambiguity. One conclusion of Bergson is that there is a unique time. By this, he means that in all the inertial frames, the rate of the time is the same. In each inertial frame, the time is a living time that people perceive through their consciousness. As an example of this distinction between real time and virtual time, Bergson brings the case of the Clock paradox (without using these words) and concludes that the traveler and the stay at home have the same age after the traveler's trip. In the appendix, he comes back to this problem with more details. The end of that chapter is devoted to Simultaneity, where Bergson distinguishes between intuitive simultaneity and simultaneity of physicists.

Chapters 5 and 6 develop the idea of the connection between time and space. The book finishes with three appendices: "The journey of the projectile", "The reciprocity of the acceleration" and "Proper time and worldline".

The book is really a work of philosophy and not very interesting for a physicist. There is no new point of view on the Theory of Relativity for a physicist. Frequently, the explanations are too long and include several repetitions. The language used by Bergson is very particular and it is very easy for a physicist to make some mistakes. In fact, his style may be irritating for a reader physicist. We shall show below that some "errors" were attributed to Bergson when a careful reading permits to understand his precise intention. And for a philosopher, a real effort is needed since Bergson does not hesitate to introduce mathematical formulas such as, for example, in the analysis of the Michelson–Morley experiment.

10.3. Debates and Critics

First, one has to say that in no place did Bergson write that the Special Theory of Relativity is wrong. Furthermore, one has to recall that he had a great esteem for Einstein as can be seen by the first words of his preface. One has to mention a very important remark of Einstein himself. During his long trip in 1922–1923 he wrote a diary and an extract is given by Canales (2005):

> Yesterday I immersed myself in the Bergson's book on relativity and time. Amazingly, he considers time but not space to be

problematic... But it seems to really understand relativity theory and not to put himself in contradiction to it.

These lines are essential for understanding Bergson and show how he was misunderstood. Even if he made errors, he did not cast doubts on the validity of the Relativity as a physical theory. And even if Einstein is quoted as saying that Bergson made errors, they are in the frame of the theory. After all, there are some textbooks that are not free of errors but they are not condemned but only corrected.

Concerning the debates provoked by different authors, the most virulent criticisms come, as expected, from physicists. From the time of the publication (D'Abro in 1923) until recently (Sokal and Bricmont), there were always researchers ready to discuss repeatedly what are his "errors". Curiously, the debate was seen as a fight between Einstein and Bergson, and naturally, one looks for the winner! (see Canales.) I have a strong suspicion that many had not read the book or had read it only partially and knew the ideas of Bergson only indirectly, a sort of second-hand knowledge.

Some examples of such a situation will be given for some formulations of Bergson. The first example is the claim of Bergson that there is only a unique time besides the plurality of times. Apparently, this contradicts the theory, in which there is no universal time. Bergson wants to say that all the clocks in all the frames are identical! It results that in all frames the rate of time is the same, the real time is lived by the observers in each inertial frame.

Another example is the word "attribute" and also the word "virtual" for an observer. When the S observer (Peter) determines the time in the S' frame (Paul), Bergson writes that this time is "attributed" by Peter to Paul. The word "attribute" has some connotation of arbitrariness as if it was a decision of Peter. But Bergson is coherent with himself. The time of Paul as seen by Peter is not a "real time" and it does not belong to anybody. Consequently, Peter "attributes" this time to Paul. The claim of Bergson is that physicists use the expression "time" for different kinds of times. But we recall that he never said that the theory is wrong, but only that he is a philosopher who writes for philosophers.

The observer in S' is "imagined" by the S observer and this may seem to be without meaning: the S' observer is also a living being! Bergson expresses by this word the fact that the two observers are not in the same kind of time; it is only for the S observer that the time is real. However, in the frame S, the time of the S' observer is only virtual which can be seen as imagined.

10.4. The Journey of the Projectile (The Clock Paradox)

The first two appendices are not completely independent, since Bergson wants to show how far reciprocity goes and how his distinction between real time and virtual time gives the possibility to contradict (correct?) some results of the Special Theory of Relativity. Here he goes against the mainstream and apparently the critics are right. In front of such a situation, two possibilities may be considered. One possibility is to say that he was completely wrong and to condemn him contemptuously. However, there is also another possibility. One has to ask why nobody succeeded in convincing him that he was wrong. Bergson was certainly not stupid, so why does the debate come to a deadlock?

As an example of the low esteem that physicists have for Bergson, one can quote a bad pun, which was made in French. A pun is almost impossible to be translated in another language. Nevertheless, I shall try to explain it. In the first appendix, Bergson does not use the word equivalent to "projectile" but "bulletgun", in French "boulet". The pun goes as follows: "le boulet de Bergson et ses boulettes". In French, "boulettes" are small balls, but in French slang, it means "huge errors".

After the first edition of the book, the French physicist Becquerel wrote to Bergson to show where was his error about the Clock paradox. The appendices were added in the second edition and in the first appendix, Bergson gives the complete letter of Becquerel that one can summarize.

Along the way of the traveler, there are clocks, which are all synchronized and during his travel, the traveler can compare the reading of his own clock with the reading of the clocks along the way. Because of the time dilation, Becquerel comes to the conclusion that after comparison of the traveler clock with the stay at home clock, the traveler is younger.

The answer of Bergson is long and cumbersome, but it is easy to summarize it: the two observers are at reciprocal positions, which are interchangeable. In other words, there is no symmetry breaking. But the story does not end with this appendix. The French physicist Merz (1923) was the first to write a book in which he intended to correct several books on Relativity including the book of Bergson. But in the absence of a reaction from Bergson, he wrote a long article in the *Revue de Philosophie* (1924) presenting the same argument as that of Becquerel: the aging difference comes from time dilation. In the same journal, Bergson gives the same answer of reciprocity. There is no reason to go on and quote other articles.

Now, one can understand why Dingle wrote a preface for *"Duration and Simultaneity"*. He finally found somebody who could understand his position. However, there is a difference: in 1965, Dingle was already denying Relativity, but Bergson did not do so.

In the second appendix, Bergson shows the reciprocity of the acceleration and this means that if one wants to introduce the accelerations in the Clock paradox, nothing changes. The reciprocity of the two observers always exists and consequently there is no aging difference.

Maybe one can propose an interpretation of the refusal of Bergson to accept the solutions of Becquerel and Merz. It is very likely that, without expressing it explicitly, he had in mind the three-frame model or the equivalent model of the frame change for the traveler when he comes back. If this is correct, one can understand the stubborn attitude of Bergson, since this model does not introduce any allusion to a symmetry breaking.

This entire story is very interesting not only because of the debate between Bergson and Einstein. About the Bergson–Einstein debate, there is considerable literature (see Canales) but we are not concerned with it.

This reveals that many scientists are not able to explain the Clock paradox. This is something analogous to the debates with Dingle and Sachs. Thinking that only the time dilation is the source of the aging difference is an error. There remain the same questions we asked all along this book: what is the role of the acceleration and the principle of equivalence?

A well-established stereotype appears very frequently in the works on Bergson, namely that Bergson did not understand the Special Theory of Relativity and made several errors. But after more than 80 years, a more balanced view has been developed and it has been suggested that the ambiguous language of some scientists may explain some errors of Bergson.

Nevertheless, so many do not see that Bergson is not necessarily wrong. For example, the acceleration is reciprocal as the relative velocity between two frames. In his 1918 article (Dialogue with Criticus), Einstein shows that the acceleration of the traveler and that of the stay at home do not have the same meaning and must be differentiated. This is the source of the symmetry breaking. We also showed the influence of the stopping of the frame because the acceleration is involved in it. It remains strange that, similarly to Dingle, nobody was able to explain to Bergson his errors in a convincing manner.

Chapter 11

The Langevin Effect

The final impression one gets after this short tour of the Clock paradox phenomenon is a failure in the understanding and interpretation of the Special Theory of Relativity. This failure is very clear when a problem (the Langevin problem) in this theory did not succeed in reaching a consensus for so long a time. The large number of solutions published during more than a century is also a clear indication of this failure.

The three debates presented above (Dingle, Sachs, and Bergson) have a common characteristic: the different actors did not reach an agreement about the age difference. At the same time, deniers of Relativity did not receive a definitive answer after their provocative pamphlet we recall at the end of Chapter 1. In other words, for a very long time, the Clock paradox was a belief and some people did not know how to solve it.

In analyzing the phenomenon by itself, one may conclude that the Clock phenomenon was "polluted" by several factors.

The first is that the large majority of scientists see only time dilation as the source of the age difference. Consequently, reciprocity, which is the basis of the problem (why the two observers do not have the same age in spite of the symmetry?) is forgotten.

Secondly, the acceleration is set aside and many people think that a short time for the acceleration is enough to forget it. In this case, the result is already "known". Consequently, the possibility for errors is low since one may be easily satisfied by a wrong solution, which seems to give a "good" answer.

Another factor is the confusion in the language and in the use of the concepts of the Special Relativity. We shall take only a few examples that we have already mentioned in the preceding chapters. One can quote sentences like "The A clock is behind the clock B and the clock B is behind the clock A" or "A moving clock goes slower". It is loose language to say something else than what is said explicitly. The dangerous side of such sentences is that some people may accept them literally. And probably there are still people who do.

The Lorentz equations suffer the same defect as loose language. Frequently, only the consequences of the equations are given without details. The length contraction is automatically applied to a moving frame (relative to another frame). It is affirmed that from the frame A at rest all the distances in the frame B moving with velocity V are seen as reduced by the factor γ.

It is a very simple and elementary calculation to compare two events in the frames S and S'. In S one has events (x_1, t_1) and (x_2, t_2). In S' one has

$$x_2' - x_1' = (1/\gamma)[(x_2 - x_1) - V(t_2 - t_1)].$$

The simple and well-known conclusion is that one cannot apply automatically the formula of length contraction.

The formula of time dilation is also applied without verification. The result is that the distinction between proper time and measured time disappears completely and one may obtain strange results. The best example is the resolution of the three-frame model in which all the frames are inertial, as given by several authors.

We recall it briefly. The frame B (the outward traveler) comes from $-\infty$, crosses the frame A, and goes on toward a star. When it reaches the star, it meets the frame C (the inward traveler) traveling with velocity $-V$ toward A. At the meeting, the frames B and C synchronize their clocks. C goes on, meets A, and disappears toward $-\infty$. The important point is that the frames are always in motion (one relative to the others) and never stop.

The relevant results are two times corresponding to two processes: (1) the elapsed time ΔT in the frame A from the departure of the frame B from the frame A, until the time of the encounter of the frame A with the frame C; (2) the time $\Delta\tau$ indicated by the clock of the frame C at the A–C encounter.

The results of the calculations by Grünbaum are as follows. The time ΔT of process 1 is L/V in the two frames A and C. Similarly, the time of process 2 is $L/\gamma V$ in the two frames A and C. The conclusion of the calculations is that the time of process 1 is the same in two frames (A and C) moving with velocity V one relative to the other. The same thing for process 2: the time of this process is the same in the two different moving frames, A and C. The time dilation does not appear! Relativity has disappeared because of the confusion between proper time and the read time.

Coming back to the fact that so many people have concluded that time dilation is enough in order to understand the paradox, this raises the question of the diffusion of knowledge. Metz, the virulent critic of Bergson, did not mention the principle of equivalence. One has the right to think that he did know it. He wrote in 1924 three articles to show the errors of Bergson without mentioning an eventual role of the acceleration. However, already several publications had appeared on this subject. The first is the 1911 article of Einstein, where he introduced this concept. Some years afterwards Einstein published his 1918 article on the Dialogue with Criticus. All these articles were published in German but maybe Metz did not know German. However, the translation of the book of Born on the Relativity had already been published in English in 1924. Furthermore, Metz did not live in the desert, and surely had contacts with other physicists. How is such a situation possible? The question becomes more acute recalling that Lorentz in his article of 1914 mentioned the acceleration and that Langevin in his solution of 1924 mentioned the gravitational field.

It is possible to ask the same question about the answers of McCrea to Dingle in 1956. At this time, the books of Tolman (1933), Møller (1952), and Born (1924) had already been published but McCrea did not mention them nor the equivalent principle. There remains the same impression: knowledge about some progress in the interpretation of the Clock paradox does not propagate linearly with time.

It is really a big surprise that some people do not know that the Clock paradox has its origin in the article of Langevin of 1911.

All that was mentioned is a clear indication that scientific knowledge is apparently spreading unequally through the scientific community. Moreover, it seems (but one has to check more carefully) that this situation is specific to the Clock paradox. The diffusion of knowledge associated with the development of a specific subject deserves to be studied attentively since in the present case, this diffusion assumes complicated and unexpected ways.

The last remark is at the pedagogical level. The Special and General Theories of Relativity are extraordinarily difficult to understand besides their intrinsic mathematical difficulties. This conclusion I drew comes not from an internal examination of the theory but how it is used. A theory formulated in the beginning of the 20th century, which received numerous experimental verifications, remains until today the center of debates and discussions. The main concept that was discussed in this book is the concept of time dilation. It is likely that it remained a misused concept for a long time. A real pedagogical effort needs to be undertaken by physicists to clear the theory of all misinterpretations. In particular, the distinction between the Lorentz version and the Einstein version of the Relativity must be clearly exposed and the consequences of each theory must be explained.

Finally, since there is no paradox, one can call the entire story the Langevin effect.

Bibliography

H. Arzeliès (1966). *Relativistic Kinematics*, Pergamon Press, Oxford.

J.S. Bell (2011). *How to Teach Relativity in Foundation of Quantum Mechanics*, University of Michigan, Ann Arbor.

B. Bensaude-Vincent (1988). When a Physicist Turns on Philosophy, Paul Langevin 1911–1939, *J. History Ideas* Vol. 49, p. 319.

H. Bergson (2009). *Durée et Simultanéité, Edition Critique*, Presses Universitaires de France, Paris.

H. Bergson (2016). *Histoire de l'idée de temps*, Presses Universitaires de France, Paris.

H. Bergson (1965). *Duration and Simultaneity*, The Bobbs-Merrill Company, Indianapolis.

H. Bergson (1924). Les temps fictifs et le temps réel, *Rev. Philos.* Vol. 31, p. 241.

D. Bohm (1965). *The Special Theory of Relativity*, Benjamin, New York.

M. Born (1924). *Einstein's Theory of Relativity*, Methuen, London, First English Translation; (1962, Second English Translation) Dover publications.

G. Builder (1957). The Resolution of the Clock Paradox, *Australian J. Phys.* Vol. 10, p. 246.

J.W. Campbell (1933). The Clock Problem in Relativity, *Phil. Mag.* Vol. 15, p. 48.

J.W Campbell (1940). The Nature of Time, *Nature* Vol. 145, p. 426.

J. Canales (2005). *The Physicist and the Philosopher: Einstein, Bergson, and the Debate that Changed Our Understanding of Time*, Princeton University Press.

H. Chang (1993). A Misunderstood Rebellion: The Twin Paradox Controversy and Herbert Dingle's Vision of Science, *Stud. Hist. Phil. Sci.* Vol. 24, p. 741.

W. Cochran (1960). The Clock Paradox, *Vistas in Astronomy* Vol. 3, p. 78.

F.S. Crawford (1957). The 'Clock Paradox' of Relativity, *Nature* Vol. 179, p. 1071.

G. Deleuze (1991). *Bergsonism*, Zone Book, New York.

H. Dingle (1965A). Relativity and Space Travel, *Nature* Vol. 177, p. 782.

H. Dingle (1965B). Relativity and Space Travel, *Nature* Vol. 178, p. 680.

H. Dingle (1965C). A Problem in Relativity Theory, *Proc. Phys. Soc. A* Vol. 69, p. 925.

H. Dingle (1957). The 'Clock Paradox' of Relativity, *Nature* Vol. 179, p. 1242.

H. Dingle (1958). Clock Paradox of Relativity, *Science* Vol. 127, p. 138.

H. Dingle (1961). *The Special Theory of Relativity*, Fourth Edition, Methuen, London.

H. Dingle (1964). Reason and Experiment in Relation to the Special Relativity Theory, *Brit. J. Phil. Sci.* Vol. 15, p. 41.

H. Dingle (1980). *The "Twins" Paradox of Relativity*, Wireless World.

A. D'Abro (1927). *Bergson ou Einstein*, Gaulon, Paris.

E. During (2014). Langevin ou le paradoxe introuvable, *Rev. Métaphysique et de Morale* Vol. 84 p. 152.

A.S. Edington (1923). *The Mathematical Theory of Relativity*, The University Press, Cambridge.

A. Einstein (1905). "Zur Elektrodynamik Bewegter Körper", *Ann. der phy.* Vol. 17, p. 891 (On the Electrodynamics of Moving Bodies).

A. Einstein (1911). Über den Einfluss der Schwercraft auf die Ausbreitung des Lichtes, *Ann. Phys.* Vol. 35, p. 898 (On the Influence of Gravitation on the Propagation of Light).

A. Einstein (1918). *Dialog über Einwände gegen die Relativitätstheorie*, Die Naturwissenschaften Vol. 6, p. 697. Published in The Collected Papers of Albert Einstein (1989) Princeton University Press, Princeton (Dialog about Objections against the Theory of Relativity).

L. Essen (1971). *The Special Theory of Relativity: A Critical Analysis*, Oxford Science Research Papers 5.

J.H. Field (2008). *The Langevin "Twin Paradox" Paper Revisited*, preprint, arXiv: 0811.3562v1.

R.M. Frye and V. Brigham (1957). Paradox of the Twins, *Amer. J. Phys.* Vol. 25, p. 553.

J. Gambon, F. Menfez, M.B. Paranjape and B. Sirois (2018). *The Twin Paradox: The Role of Acceleration*, preprint, arXiv: 1807.02148v1.

G. Gale (2019). *Cosmology: Methodological Debates in the 1930s and 1940s*, Summer Edition, *The Stanford Encyclopedia of Philosophy* Edward N. Zalta (ed.), https://plato.stanford.edu/archives/sum2019/entries/cosmology-30s/.

R.H. Good (1982). Uniformly Accelerated Reference Frame and Twin Paradox, *Amer. J. Phys.* Vol. 50, p. 232.

D.M. Greenberger (1972). The Reality of the Twin Paradox Effect, *Amer. J. Phys.* Vol. 40, p. 751.

A. Grünbaum (1954). The Clock Paradox in the Special Theory of Relativity, *Philos. Sci.* Vol. 21, p. 249.

J.C. Hafele and R.E. Keating (1972A). Around-the-World Atomic Clocks: Predicted Relativistic Time Gains, *Science* Vol. 177, p. 166.

J.C. Hafele and R.E. Keating (1972B). Around-the-World Atomic Clocks: Observed Relativistic Time Gains, *Science* Vol. 177, p. 168.

L. Iorio (2005). An Analytic Treatment of the Clock Paradox in the Framework of the Special and General Theories of Relativity, *Found. Phys. Lett.* Vol. 18, p. 1.

A. Kopff (1923). *The Mathematical Theory of Relativity*, Methuen, London.

L.D. Landau and E.M. Lifshitz (2013). *The Classical Theory of Fields*, Elsevier Science Ltd, Oxford.

P. Langevin (1911). *L'Evolution de l'espace et du temps*, Scientia Vol. 10, p. 31. The following URL gives a copy of the original article in the journal Scientia: https://amshistorica.unibo.it/diglib.php?inv=7&int_ptnum=10&term_ptnum=39&format=jpg; Online fr.wikisource.org (in French); cn.wikisource.org (in English).

H. Lass (1963). Accelerating Frames of Reference and the Clock Paradox, *Amer. J. Phys.* Vol. 31, p. 274.

C.B. Leffert and T.M. Donahue (1958) Clock Paradox and the Physics of Discontinuous Gravitational Fields, *Amer. J. Phys.* Vol. 26, pp. 515–523.

J.C. Levy-Leblond (2007). Bergson et la Science, *Ann. Bergsoniennes* Vol. 3, p. 237.

H.A. Lorentz (1914). Considérations élémentaires sur le principe of relativité, *Rev Générale Sci.* Vol. 15, p. 179.

L. Marder (1971). *Time and the Space — Traveler*, University of Pennsylvania Press, Philadelphia.

W.H. McCrea (1951). The Clock Paradox in Relativity Theory, *Nature* Vol. 167, p. 680.

W.H. McCrea (1956). Answer to Dingle, *Nature* Vol. 177, p. 784.

E.M. McMillan (1957) "The Clock Paradox" and Space Travel, *Science* Vol. 128, p. 381.

A. Metz (1923). *La Relativité: exposé sans formules des théories d'Einstein et réfutation des erreurs contenues dans les ouvrages les plus notoires*, Chiron Editions, Paris.

A. Metz (1924). Le temps d'Einstein et de la philosophie, *Rev. Philos.* Vol. 31, p. 56.

E. Minguzzi (2004). *Differential Aging from Acceleration: An Explicit Formula*, preprint, arXiv:physics/0411233v1.

M. Morand (1922). Einstein au Collège de France, *Nature* Vol. 1, p. 315.

C. Møller (1952). *The Theory of Relativity*, Clarendon Press, Oxford.

R.A. Muller (1972). The Twin Paradox in Special Relativity, *Amer. J. Phys.* Vol. 40, p. 966.

T. Müller, A. King and D. Adis (2008). A Trip to the End of the Universe and the Twin Paradox, *Amer. J. Phys.* Vol. 76, p. 360.

C. Nordman (2011). *Einstein Presents and Discusses His Theory*, 21st Century Science and Technology, Summer, p. 9.

R. Perrin (1979). Twin Paradox: A Complete Treatment from the Point of View of Each Twin, *Amer. J. Phys.* Vol. 47, p. 317.

P. Pesic (2003). Einstein and the Twin Paradox, *Eur. J. Phys.* Vol. 24, p. 585.

W.A. Rodrigues and M.A.F. Rosa (1989). The Meaning of Time in the Theory of Relativity and "Einstein's Later View of the Twin Paradox", *Found. Phys.* Vol. 19, p. 705.

C.A. Ronan (1979). Obituary: Herbert Dingle 1890–1978, *J. Brit. Astron. Assoc.* Vol. 89, p. 288.

M. Sachs (1971). *A Resolution of the Clock Paradox*, Phys. Today p. 23.

M. Sachs (1974). On Dingle's Controversy about the Clock Paradox and the Evolution of Ideas in Science, *Int. J. Theor. Phys.* Vol. 10, p. 321.

G.D. Scott (1959). On Solutions of the Clock Paradox, *Amer. J. Phys.* Vol. 27, p. 580.

Société française de philosophie (1922). Grandes conférences en téléchargement, No. 13, Albert Einstein, (www.sofrphilo.fr).

A. Sokal and J. Bricmont (1998). *Intellectual Impostures: Postmodern Philosophers' Abuse of Science*, Profile Books, Colchester, UK.

C. Semay (2006). Observer with a constant proper acceleration, *Eur. J. Phys.* Vol. 27, p. 1157.

L. Strauton and H. van Dam (1980). Graphical Introduction to the Special Theory of Relativity, *Amer. J. Phys.* Vol. 49, p. 807.

R.C. Tolman (1934). *Relativity, Thermodynamics and Cosmology*, Clarendon Press, Oxford.

M.A. Tonnelat (1959). *Les principes de la théorie électromagnétique et de la relativité*, Masson, Paris.

C.S. Unnikrishnan (2005). On Einstein's Resolution of the Twin Paradox, *Current Sci.* Vol. 89, p. 2009.

M. von Laue (1912). Two Objections Against the Theory of Relativity and their Refutation, *Phys. Zeit.* Vol. 13, p. 118. (Translation in Wikisource).

H. Weyl (1922). *Space–Time–Matter*, Methuen, London.

T.-Y. Wu and Y.C. Lee (1972). The Clock Paradox in the Relativity Theory, *Int. J. Theor. Phys.* Vol. 5, p. 307.

Index